KATE HALE

Growth Spurts and Physical Development

What to Expect During Key Growth Phases

Copyright © 2024 by KATE HALE

All rights reserved. No part of this publication may be reproduced, stored or transmitted in any form or by any means, electronic, mechanical, photocopying, recording, scanning, or otherwise without written permission from the publisher. It is illegal to copy this book, post it to a website, or distribute it by any other means without permission.

First edition

*This book was professionally typeset on Reedsy.
Find out more at reedsy.com*

Contents

Introduction	1
Infant Growth (0-12 Months)	7
Toddler Growth (1-3 Years)	15
Preschool Growth (3-5 Years)	23
School-Age Growth Spurts (6-9 Years)	32
Puberty and Its Physical Impact (13-15 Years	41
Late Adolescence (16-18 Years)	50
Nutrition for Optimal Growth	59
Genetics and Growth	68
Sleep and Growth	77
Signs of Growth Delay	86
Medical Conditions Affecting Growth	96
When to See a Doctor	106
Conclusion	114

Introduction

Growth spurts are crucial phases in a child's development, characterized by rapid changes in height, weight, and overall physical structure. These periods of accelerated growth typically occur at specific stages during childhood and adolescence, although their timing and intensity can vary among individuals. Understanding growth spurts is essential for parents and caregivers, as these moments signify more than just physical expansion; they are indicators of healthy development across multiple domains, including cognitive and emotional growth.

What Are Growth Spurts?

A growth spurt is defined as a period of rapid physical growth, marked by significant changes in height and weight over a short period. During these spurts, children often experience heightened hunger, fatigue, and a noticeable increase in appetite. This phase of rapid growth usually coincides with hormonal changes that influence not only physical dimensions but also a child's mental and emotional state. While growth occurs throughout childhood, it's in these spurts that the most dramatic changes are observed, with children sometimes growing several inches in a few months.

Growth spurts are closely linked to the endocrine system, particularly the release of growth hormones, which are regulated by the pituitary gland. These hormones stimulate growth in various parts of the body, including bones, muscles, and other tissues. This is why growth spurts can feel sudden and unpredictable, as they depend on the body's internal hormonal rhythms. For instance, newborns experience multiple spurts during their first year of life, while adolescents undergo the most dramatic and prolonged growth spurts during puberty.

One notable feature of growth spurts is their asymmetry. Some children experience rapid growth in height first, followed by an increase in weight, or vice versa. This uneven development can sometimes cause temporary coordination issues or a period of awkwardness as the body adjusts to its new size and shape. Parents may notice their children becoming clumsier or experiencing discomfort in their joints and muscles as their body rapidly expands.

Physical Development and Its Phases

Physical development unfolds in distinct phases, with growth spurts punctuating these stages. Each phase is marked by specific milestones that reflect the body's increasing complexity and capabilities. These phases of development typically align with a child's age but can vary based on individual genetics, nutrition, and environmental factors.

Infancy (0-2 years): The first phase of rapid physical development occurs during infancy. From birth to about two years, babies grow at an astonishing rate, doubling their birth weight by five months and tripling it by the first year. During this time, physical development focuses on foundational motor skills, such as lifting the head, rolling over, and sitting up. Growth spurts during infancy can be frequent, leading to dramatic changes in size and behavior within weeks. This stage also sets the foundation for more complex motor skills, such as crawling, standing, and walking.

Early Childhood (2-6 years): Following infancy, children enter a phase of more gradual growth. However, the body continues to build on the foundation laid during the earlier years. During early childhood, children

INTRODUCTION

typically grow between 2.5 to 3 inches per year and gain about 4-6 pounds annually. The development of fine motor skills becomes a priority during this stage, as children learn to manipulate small objects, draw, and use tools like utensils. Growth spurts in early childhood may not be as dramatic as those seen in infancy or adolescence, but they are still crucial for laying the groundwork for future physical and cognitive development.

Middle Childhood (6-12 years): Growth during middle childhood is relatively steady but punctuated by periods of accelerated development. Children in this phase grow about 2 inches per year until they reach puberty. During this stage, muscle mass increases, and children begin to refine their gross and fine motor skills. Growth spurts in middle childhood often coincide with a significant increase in physical activity, which supports the development of muscle tone and coordination. Additionally, this stage is when children begin to develop more complex motor patterns, such as those required for sports or dance.

Adolescence (12-18 years): The most dramatic and prolonged growth spurts occur during adolescence, triggered by the onset of puberty. In this phase, the body undergoes rapid physical changes driven by a surge in hormones such as testosterone and estrogen. Boys typically experience their growth spurt later than girls, but they tend to grow taller and build more muscle mass. Girls, on the other hand, experience growth spurts earlier, usually between the ages of 10 and 14, and their development focuses more on the accumulation of fat in areas such as the hips and breasts. During adolescence, children may grow several inches in a single year, and their weight gain can double as their bodies adjust to adult proportions. Emotional and cognitive development also accelerates during this phase, making it a critical time for overall maturation.

The Importance of Tracking Milestones

Monitoring a child's growth through milestones is an essential aspect of ensuring healthy development. Growth milestones are indicators of a child's overall health and can help identify potential developmental delays or health issues. While every child grows at their own pace, there are specific markers that parents and pediatricians use to assess whether a child is progressing within a healthy range.

Growth milestones are often tracked using growth charts, which plot a child's height, weight, and head circumference against standardized averages for their age and gender. These charts provide a visual representation of how a child compares to others in their peer group and can help doctors and parents identify any abnormalities in growth. For instance, a sudden deviation from a child's established growth pattern may signal an underlying health issue, such as a hormonal imbalance, nutritional deficiency, or chronic illness.

In addition to physical markers like height and weight, milestones also encompass motor skills, cognitive abilities, and social-emotional development. For example, a two-year-old should be able to walk independently, use simple phrases, and show interest in playing with others. If a child is not meeting these developmental milestones, it may indicate a delay that requires further evaluation.

Tracking milestones is also important for recognizing the emotional and psychological aspects of growth spurts. Children may experience mood swings, irritability, and changes in behavior during these periods of rapid physical development. By being aware of when growth spurts are likely to occur, parents can better support their children through these transitions, providing reassurance and guidance as their bodies change.

It's important to note that while growth spurts can sometimes seem sudden, they are part of a continuous developmental process. Children grow in spurts, but the groundwork for these bursts of growth is laid gradually over time. Therefore, tracking a child's overall growth trajectory is more important than focusing on any single spurt. Regular check-ups with a pediatrician are

INTRODUCTION

crucial for ensuring that children are meeting their growth milestones and receiving the necessary support for their physical, emotional, and cognitive development.

Common Myths about Growth

Despite the wealth of information available on child development, several myths and misconceptions persist regarding growth spurts. These myths can lead to unnecessary worry or confusion among parents, making it essential to dispel them with evidence-based facts.

Myth 1: All Children Grow at the Same Rate

One of the most common misconceptions is that all children grow at a uniform pace. In reality, growth rates can vary significantly between children, even among siblings. Factors such as genetics, nutrition, and overall health play a critical role in determining how and when a child experiences growth spurts. It's entirely normal for some children to grow rapidly during certain phases and then slow down, while others may experience more gradual, consistent growth.

Myth 2: Height Predicts Future Success

Another persistent myth is that taller children are destined for greater success in life. While height can influence certain social dynamics, there is no direct correlation between height and success in education, career, or personal fulfillment. Emotional intelligence, perseverance, and social skills are far more important indicators of future success than a child's physical stature.

Myth 3: Growth Spurts Cause Fevers

Some parents believe that growth spurts are accompanied by fevers or other signs of illness. While growth spurts can cause discomfort, such as muscle aches or growing pains, they are not associated with fevers or infections. If a child develops a fever during a growth spurt, it's likely due to an unrelated illness, and medical attention may be required.

Myth 4: Boys Always Grow Taller Than Girls

While it is true that, on average, boys tend to grow taller than girls, this is not a hard and fast rule. Many girls grow taller than their male peers during childhood, particularly because they experience growth spurts earlier in life. Additionally, adult height is influenced by a complex interplay of genetics, nutrition, and environmental factors, meaning there are many exceptions to this general trend.

Myth 5: Puberty Ends Growth Spurts

Many people assume that once puberty is complete, growth stops. While it is true that most growth occurs during puberty, some individuals continue to grow into their early twenties, especially in terms of muscle mass and bone density. Boys, in particular, may experience muscle development and changes in body composition well after their final height has been reached.

Final Note

Understanding growth spurts and the physical development phases that children go through can significantly reduce the anxiety associated with raising children. By tracking milestones and dispelling common myths, parents can support their children's growth in a way that promotes both physical and emotional health. Growth is not a straightforward process; it is a dynamic and complex journey that varies from one individual to another. Recognizing the importance of each growth phase allows for a more supportive and informed approach to parenting, ultimately leading to healthier, happier children.

Infant Growth (0-12 Months)

During the first year of life, an infant undergoes the most rapid and remarkable physical development compared to any other period in human life. The changes that occur within the first twelve months are essential for setting the stage for further growth and development. During this phase, newborns develop foundational skills and experience significant progress in height, weight, and motor abilities. Each aspect of infant growth is interconnected, emphasizing the importance of proper care and nutrition during this critical time.

Newborn Physical Changes

At birth, newborns typically weigh between 5.5 and 8.8 pounds and measure about 18 to 22 inches in length. However, these numbers vary depending on genetic factors, maternal health, and the duration of the pregnancy. Newborns undergo a range of physical changes in the early days and weeks, starting with significant adaptations to life outside the womb. Their bodies shift from relying on maternal nourishment through the placenta to processing nutrition independently through breastfeeding or formula. This transition is accompanied by changes in their respiratory, circulatory, and digestive systems.

In the first few days, newborns may lose up to 10% of their birth weight due to the expulsion of fluids and the adjustment to feeding. This weight loss is normal and expected, but by the second week, most babies regain their birth weight and continue to grow rapidly. The skin, which may appear wrinkled and soft at birth, begins to smooth out as the baby gains fat. The fontanelles, or soft spots on the baby's head, allow the skull to remain flexible during delivery. These soft spots gradually close as the baby's skull hardens over the next 18 months.

Newborn reflexes also serve as indicators of healthy development. Reflexes like rooting (turning the head toward a touch on the cheek to find food), sucking, grasping, and the Moro reflex (startle reflex) are essential for survival and indicate proper neurological functioning. As the months progress, these reflexes will give way to more voluntary movements, which are essential in the development of motor skills.

During the first few months, newborns display what is known as cephalocaudal development—growth that progresses from the head down to the rest of the body. For instance, a newborn gains control of the neck muscles before developing control of the arms and legs. This pattern of growth is vital in the later acquisition of skills such as sitting, crawling, and walking. Physical changes during this time are not just visible in growth charts but are also represented in an infant's improved coordination and ability to interact with their surroundings.

Height and Weight Milestones

Monitoring an infant's growth through height and weight milestones is one of the most effective ways to assess their development and ensure they are growing at a healthy pace. Pediatricians commonly use growth charts, which compare a child's growth to standardized norms for their age and sex. These charts help parents and healthcare providers track an infant's progress and spot potential health concerns early.

By three months, most infants have already gained significant weight and length since birth. An average infant will gain about 5 to 7 ounces each week during the first six months, leading to a doubling of their birth weight by the time they are five to six months old. In terms of height, babies grow about 1 to 1.5 inches per month during the first six months. This rapid pace slows slightly in the second half of the first year, but growth remains significant as the baby continues to gain about 3 to 5 ounces per week and grows 0.5 inches per month from six to twelve months. By the end of the first year, most babies have tripled their birth weight and grown about 10 inches.

Weight gain during this period is essential not only for healthy development but also for providing the energy required for the infant's increasing activity levels. Fat stores accumulated during this time serve as an energy reserve and help with temperature regulation. While babies grow rapidly during their first year, the rate of growth can vary between individuals, and genetics play a large role in determining an infant's ultimate growth trajectory.

It is important to note that growth spurts are common during the first year. These spurts, which often occur around 3 weeks, 6 weeks, 3 months, and 6 months, can lead to sudden increases in appetite and fussiness, signaling that the baby needs more nutrition to fuel their growth. These periods are also when parents may notice that their baby suddenly outgrows clothes or seems heavier when picked up.

Although every baby grows at their own rate, falling behind in growth can signal potential problems such as nutritional deficiencies or underlying medical issues. Regular pediatric check-ups are crucial for monitoring growth and ensuring that babies are progressing as expected.

Cognitive and Motor Skill Development

Cognitive and motor skill development begins in infancy and progresses rapidly during the first year. Cognitive development refers to the infant's ability to learn, think, and process information, while motor development focuses on their ability to move and manipulate their environment. These two domains of development are closely intertwined, as improved motor skills often lead to increased opportunities for learning and exploration.

At birth, a newborn's brain is about 25% of its adult size, but it grows rapidly in the first year as new neural connections are formed. This brain growth supports the development of cognitive skills, such as recognizing faces, responding to voices, and beginning to understand cause-and-effect relationships. For example, a baby might learn that crying will bring attention from a caregiver or that kicking a toy will make it move.

In the early months, infants rely heavily on their senses to understand the world around them. Newborns can see objects that are 8 to 12 inches away, making them particularly attuned to faces. By 2 to 3 months, babies begin to smile in response to social stimuli, a milestone that indicates both cognitive and emotional development. At around 4 to 6 months, babies start to show more interest in their surroundings, using their hands and mouths to explore objects. This sensory exploration is crucial for brain development, as it helps babies learn about different textures, shapes, and sounds.

Motor skill development can be divided into two categories: gross motor skills and fine motor skills. Gross motor skills involve large movements, such as rolling over, sitting up, and eventually walking, while fine motor skills involve smaller, more precise movements, such as grasping objects or picking up small items.

During the first three months, babies primarily focus on developing gross motor skills by gaining control over their head and neck muscles. By three months, most infants can lift their heads when lying on their stomachs, a key milestone in the development of other gross motor skills like sitting and crawling. Fine motor skills also begin to develop during this time, as babies start to open and close their hands and bring their hands to their mouths.

Between four and six months, babies gain more control over their bodies.

Many babies learn to roll over from their stomach to their back during this time, and some may begin to sit up with support. Fine motor skills continue to improve as babies learn to reach for and grasp objects, often transferring them from one hand to the other. This period of exploration is crucial for cognitive development, as it allows babies to interact with and learn from their environment.

By the time babies reach six to nine months, their gross motor skills have advanced significantly. Most babies can sit without support, and some may begin to crawl or scoot. During this stage, babies also start to develop the strength and coordination needed to pull themselves up to a standing position. Fine motor skills continue to improve, with babies becoming more adept at picking up small objects using a pincer grasp (using the thumb and forefinger).

Between nine and twelve months, many babies take their first steps, marking a major milestone in gross motor development. While some infants may walk earlier or later, this period is typically when babies begin to experiment with standing and walking. Fine motor skills also continue to improve, with babies becoming more skilled at manipulating objects, such as stacking blocks or feeding themselves finger foods.

Cognitive development also accelerates during the second half of the first year. Babies begin to understand object permanence—the concept that objects continue to exist even when they are out of sight. This newfound understanding can lead to separation anxiety, as babies become more aware of their caregivers' presence and absence. Language development also begins to emerge during this time, with many babies babbling and experimenting with sounds. While true words may not yet be formed, the foundations of language are being built as babies listen to and mimic the sounds they hear.

How Nutrition Shapes Early Growth

Nutrition plays a fundamental role in shaping an infant's growth during the first year of life. Proper nutrition provides the essential building blocks for physical, cognitive, and motor development. Infants require a diet rich in nutrients such as proteins, fats, vitamins, and minerals to support their rapid growth and brain development.

For most infants, breastfeeding or formula provides all the necessary nutrients during the first six months of life. Breast milk is considered the gold standard for infant nutrition, as it contains the perfect balance of nutrients and antibodies that help protect against infections and illnesses. Breastfed babies typically receive all the essential nutrients they need, including fats that support brain development and growth hormones that promote healthy weight gain.

Formula is a suitable alternative for babies who cannot be breastfed or whose mothers choose not to breastfeed. Modern formulas are designed to mimic the nutritional profile of breast milk as closely as possible, providing the necessary calories, proteins, and fats for healthy growth.

Around six months, most babies are ready to begin solid foods. This transition is important for meeting the increasing nutritional needs of growing infants. Iron, in particular, becomes critical at this stage, as babies' iron stores begin to deplete after six months. Introducing iron-rich foods, such as fortified cereals, meats, and leafy greens, helps prevent iron deficiency, which can impact cognitive development and overall health.

As babies begin to eat solid foods, it's important to offer a variety of nutrient-dense options. Vegetables, fruits, grains, and proteins should all be introduced gradually to ensure that babies receive a balanced array of vitamins and minerals that support their development. Early exposure to a wide variety of foods also sets the foundation for healthy eating habits later in life. During this time, parents are encouraged to offer soft, mashed, or pureed foods that are easy for babies to swallow and digest.

Fats remain an essential component of an infant's diet throughout the first year, as they are critical for brain development. The human brain grows rapidly during the first two years of life, and fats, particularly those containing omega-3 fatty acids like DHA (docosahexaenoic acid), are vital for the development of neural connections. Breast milk naturally contains high levels of these essential fats, while formulas are fortified with DHA to mimic the benefits of breast milk. As infants transition to solid foods, it's important to continue offering healthy sources of fats, such as avocado, yogurt, and fatty fish, to support ongoing brain development.

Hydration is another key component of nutrition in the first year. For breastfed and formula-fed infants, milk provides adequate hydration during the first six months. Once solids are introduced, water can be offered in small amounts to complement feedings, particularly as babies become more mobile and active. However, milk (either breast milk or formula) should remain the primary source of nutrition until at least 12 months of age.

While solid foods play an increasingly important role in nutrition after six months, the transition should be gradual. Parents are often advised to introduce one new food at a time, waiting a few days between each new item to monitor for any allergic reactions. Common allergenic foods such as eggs, nuts, and fish can be introduced early under the guidance of a pediatrician, as research suggests that early exposure may reduce the risk of developing food allergies.

As babies approach their first birthday, their nutritional needs continue to evolve. They begin to eat more solid foods, gradually reducing their reliance on breast milk or formula. By 12 months, most babies are able to eat a wide variety of family foods, albeit in smaller, softer portions. Ensuring that infants are exposed to a diverse range of nutrients during this period is crucial for supporting ongoing physical growth, cognitive development, and immune system health.

It is important to remember that nutritional needs vary from baby to baby. Some infants may have difficulty gaining weight or may experience feeding challenges, such as food aversions or sensitivities. In these cases, parents should consult with a pediatrician or a pediatric nutritionist to ensure their baby is receiving adequate nutrition and to address any potential underlying health issues. Special attention may be required for premature babies, who often have different nutritional needs and growth patterns compared to full-term infants.

Finally, establishing healthy feeding habits during infancy sets the stage for lifelong wellness. Responsive feeding, where caregivers offer food in response to the baby's hunger and satiety cues, helps prevent overeating and encourages the development of self-regulation. By providing a variety of nutrient-dense foods, avoiding added sugars and overly processed items, and ensuring that mealtimes are positive and stress-free, parents can help their babies develop a healthy relationship with food that will benefit them throughout childhood and beyond.

In summary, the first year of life is a time of unparalleled growth and development, with nutrition playing a central role in shaping an infant's physical and cognitive progress. Newborns undergo remarkable changes in height, weight, and motor skills, and each of these milestones is supported by the proper intake of essential nutrients. From breastfeeding or formula feeding in the early months to the introduction of solid foods at around six months, nutrition remains the foundation for healthy development. By understanding the critical importance of early nutrition and making informed choices, parents can help their babies thrive and lay the groundwork for continued growth and well-being in the years to come.

Toddler Growth (1-3 Years)

Between the ages of one and three, toddlers experience significant developmental changes across multiple domains, including physical, cognitive, and emotional growth. While the rapid growth phase that characterized infancy begins to slow down during these years, toddlers continue to make steady and noticeable progress in height, weight, and motor skills. This period is marked by the development of more sophisticated language abilities, increased muscle strength, and improved coordination. Additionally, toddlers may experience occasional discomfort due to growing pains, a common phenomenon during this phase. Understanding these changes helps caregivers provide the support toddlers need to navigate this crucial stage of growth.

The Rapid Growth Phase

Although the growth rate in toddlers is not as intense as during infancy, it is still substantial. In the first year of life, babies grow at a remarkable rate, doubling or tripling their birth weight and rapidly increasing in length. Between one and three years old, the rate of growth slows down, but it remains consistent and steady. On average, toddlers gain about 4 to 6 pounds per year and grow approximately 3 to 4 inches annually.

During this phase, the proportions of a toddler's body begin to change. While babies tend to have larger heads in comparison to their bodies, toddlers begin to even out in proportions as the torso, legs, and arms grow. This shift in proportions marks a transition from the infant phase to a more upright, balanced posture typical of early childhood. As toddlers gain height and weight, they also lose some of the baby fat that helped regulate their temperature and energy stores in infancy. Their bodies start to slim down, and their muscle mass increases to support more active, mobile behavior.

Hormonal changes, driven by the pituitary gland, continue to influence growth during this time. Growth hormones, thyroid hormones, and insulin-like growth factors all work in tandem to promote the development of bones and tissues. Additionally, genetics plays a critical role in determining a toddler's growth trajectory. While the first year of life is largely dictated by early nutrition and prenatal conditions, the toddler years begin to reflect more of the genetic blueprint inherited from parents. Some toddlers may experience growth spurts at different times, which can lead to periods of heightened appetite and moodiness as their bodies adjust to new heights and weights.

Nutrition remains a key factor in supporting growth during this phase. Toddlers need a balanced diet rich in essential vitamins and minerals to ensure their bodies develop optimally. Calcium, vitamin D, protein, and healthy fats are especially important for building strong bones and muscles. However, growth can vary widely among toddlers due to individual differences in metabolism, activity levels, and genetic factors. While some toddlers may appear to grow rapidly, others may progress at a slower but steady pace. Regular pediatric check-ups help ensure that a toddler's growth is on track, with height and weight measurements plotted on standardized growth charts to monitor development.

One of the most noticeable changes in a toddler's appearance during this phase is the shift from a baby-like roundness to a leaner, more muscular build. This is especially evident in the legs and arms, as toddlers become more active and engage in physical play. As toddlers continue to grow, their need for sleep remains critical, as growth hormones are released primarily during deep sleep. Toddlers who do not get enough rest may experience disruptions in their growth patterns, highlighting the importance of establishing healthy sleep routines.

Speech and Language Development

Language development is one of the most remarkable and rapid areas of growth during the toddler years. Between the ages of one and three, toddlers move from using single words to forming more complex sentences, allowing them to communicate their needs, thoughts, and feelings more effectively. Speech and language development during this time is shaped by both biological factors, such as brain maturation, and environmental influences, including interactions with caregivers and exposure to language.

By the time toddlers reach their first birthday, many are already using a handful of words, such as "mama," "dada," or "ball." These early words are often related to the people and objects most familiar to them. Around 18 months, toddlers typically experience a language explosion, during which their vocabulary rapidly expands. They begin to learn new words at an astonishing rate, often acquiring several new words each week. By the time they reach two years of age, most toddlers can use between 200 and 300 words and are able to combine two or more words to form simple sentences, such as "want juice" or "big truck."

This rapid development of language is closely linked to cognitive growth. As toddlers' brains mature, they become more adept at understanding and producing language. At the same time, toddlers' growing cognitive abilities allow them to grasp more complex concepts, such as cause and effect, object permanence, and the ability to categorize objects by type, size, or color. These cognitive skills enhance toddlers' ability to use language to describe the world around them and communicate their thoughts more effectively.

Language development during this period is not just about speaking; it also involves understanding and processing language. Toddlers are able to comprehend far more words than they can say, which is why they can follow simple instructions, such as "bring me the ball" or "put your shoes on." This ability to understand language, known as receptive language, typically develops ahead of expressive language, which involves producing words and sentences.

Parents and caregivers play a crucial role in supporting speech and language development during the toddler years. Toddlers learn language through social interaction, and caregivers who engage in frequent, responsive communication with their children help stimulate language growth. Simple activities, such as reading books, singing songs, and engaging in conversations, provide valuable opportunities for toddlers to hear and practice new words. Caregivers can also help by labeling objects, describing actions, and expanding on toddlers' verbal attempts, turning a single word like "dog" into a more complete sentence like "Yes, that's a big, brown dog."

Despite the rapid progress in language development, there can be significant variability in the timing and pace of speech milestones. Some toddlers may begin speaking earlier, while others may take longer to form sentences. Bilingual toddlers, for example, may start speaking slightly later as they process multiple languages, but they often catch up quickly and benefit from the cognitive advantages of bilingualism. While it's normal for language development to vary, significant delays in speech or difficulty understanding language may signal the need for early intervention.

Muscle Strengthening and Coordination

During the toddler years, physical growth is accompanied by significant improvements in muscle strength and coordination. As toddlers become more active and mobile, they develop the strength and agility needed to perform increasingly complex movements, such as walking, running, climbing, and jumping. These improvements in gross motor skills are

TODDLER GROWTH (1-3 YEARS)

essential for exploring their environment and gaining independence.

By the time toddlers reach their first birthday, many have already taken their first steps. Walking is a major milestone in gross motor development, marking the transition from infancy to early childhood. As toddlers become more confident walkers, they quickly move on to more advanced gross motor skills, such as running, climbing stairs, and kicking a ball. These skills require not only muscle strength but also balance and coordination, which improve as the nervous system matures and the brain develops better control over the body.

Muscle strengthening during this phase is driven largely by increased physical activity. Toddlers are naturally curious and energetic, and they spend much of their time exploring their surroundings, which helps build the muscles needed for movement. Activities such as crawling, walking, climbing, and playing with toys that require pushing or pulling help toddlers develop both upper and lower body strength.

In addition to gross motor skills, toddlers also make significant progress in fine motor skills, which involve smaller, more precise movements of the hands and fingers. By the age of two, most toddlers can stack blocks, turn the pages of a book, and use a spoon to feed themselves. These activities require the coordination of multiple muscle groups, as well as hand-eye coordination, which improves steadily during this period.

Play is a critical part of physical development during the toddler years. Whether it's playing with balls, climbing on playground equipment, or participating in simple games like "Simon Says," physical play helps toddlers build muscle strength, improve coordination, and develop spatial awareness. Providing a safe environment for active play, both indoors and outdoors, encourages toddlers to move their bodies and practice new skills.

By the time toddlers reach three years old, many are able to run, jump, pedal a tricycle, and even attempt to catch a ball. These activities require coordination between the arms, legs, and core muscles, and they represent significant progress in both gross and fine motor development. Muscle tone and strength continue to improve as toddlers become more active, and their increasing physical capabilities allow them to engage in more complex and varied forms of play.

Dealing with Toddler Growth Pains

As toddlers continue to grow and develop, they may occasionally experience discomfort known as growth pains. Growth pains are common during periods of rapid physical development and are characterized by aching or throbbing sensations in the legs, particularly in the shins, calves, or thighs. While the exact cause of growth pains is not fully understood, they are thought to be related to the stretching and expansion of muscles and tendons as the bones grow.

Growth pains typically occur in the late afternoon or evening and may even wake toddlers from sleep at night. They are usually not accompanied by swelling, redness, or tenderness, which distinguishes them from injuries or other medical conditions. Growth pains tend to come and go, often lasting for a few days at a time before subsiding.

For many toddlers, growth pains are a temporary and manageable part of development. Caregivers can help alleviate discomfort by gently massaging the affected areas, applying a warm compress, or encouraging stretching exercises to loosen tight muscles. In some cases, over-the-counter pain relievers, such as acetaminophen or ibuprofen, may be recommended to ease the pain. Ensuring that toddlers get plenty of rest and sleep is also important, as growth hormones are released primarily during deep sleep.

While growth pains are typically harmless, they can be distressing for toddlers who may not yet have the language skills to express their discomfort clearly.

Caregivers should be attentive to their toddlers' complaints and provide reassurance, helping them understand that the pain is a normal part of growing. In most cases, growth pains subside on their own, and they do not lead to long-term issues or interfere with a toddler's ability to engage in physical activities.

However, it's important to recognize the difference between typical growth pains and signs of more serious medical conditions. If a toddler's pain is persistent, severe, or accompanied by other symptoms such as swelling, limping, fever, or a loss of appetite, it's crucial to consult a pediatrician. These signs may indicate an underlying issue, such as an injury, infection, or orthopedic condition that requires medical attention.

Parents can help prevent some discomfort associated with growing by ensuring their toddlers maintain an active and balanced lifestyle. Regular physical activity strengthens muscles and improves flexibility, which can reduce the strain on growing bones and joints. Additionally, maintaining a balanced diet rich in calcium, vitamin D, and other essential nutrients helps promote healthy bone development, which may minimize the frequency and intensity of growth pains.

In conclusion, growth pains are a common, albeit sometimes uncomfortable, part of the toddler years. With the right approach—balancing physical activity, nutrition, and proper rest—most toddlers can navigate this phase with minimal disruption to their daily lives. Parents play a key role in offering comfort and ensuring their child's physical and emotional well-being during these periods of rapid growth and change.

s

The toddler years are a dynamic period of growth and development, with significant milestones achieved in physical stature, cognitive abilities, and emotional regulation. During this time, toddlers experience a steady progression in height and weight, language acquisition, and muscle strength, all of which contribute to their increasing independence and engagement with the world. Growth during these years is influenced by a combination of genetics, nutrition, physical activity, and environmental factors, and it is essential for caregivers to provide a supportive, nurturing environment to foster healthy development.

Understanding the phases of toddler growth, along with the challenges like growth pains, helps parents better anticipate their child's needs and respond appropriately. Whether through encouraging active play, supporting language development, or addressing discomfort from growing pains, caregivers can actively contribute to a toddler's physical and emotional well-being. By the time children transition out of the toddler years, they will have laid a strong foundation for future growth and development, both physically and cognitively, preparing them for the next stages of childhood.

These early years are not only critical for physical development but also for forming the bonds and interactions that shape a child's emotional and psychological growth. As toddlers explore the world around them, they also learn how to navigate relationships, express themselves, and build confidence, all essential elements for healthy development in the years to come. With the right support, toddlers can thrive during this pivotal stage of their lives, setting the stage for continued growth and achievement.

Preschool Growth (3-5 Years)

Between the ages of three and five, children undergo a steady, yet significant transformation in physical, cognitive, and emotional development. While the rapid growth spurts of infancy and toddler hood begin to level out, the preschool years are marked by a more gradual but consistent progression in height, weight, and overall physical capabilities. This period is also critical for emotional and social development, as children start to engage more deeply with their surroundings and form relationships outside the family unit. Identifying and addressing any potential delays during this phase is crucial for ensuring that children meet important developmental milestones. Additionally, fostering healthy habits around nutrition and physical activity sets the stage for lifelong well-being.

Steady Growth and Physical Changes

During the preschool years, growth continues at a steady pace, though it is noticeably slower than the rapid increases seen in infancy and toddler hood. On average, children grow about 2.5 to 3.5 inches in height per year and gain around 4 to 5 pounds annually. This consistent growth helps preschoolers develop more proportionate and leaner bodies as they lose the baby fat that characterized their earlier years. By the time children reach five years old, many have a more elongated appearance, with limbs growing longer in relation to their torso, and their heads becoming more proportionate to their bodies.

GROWTH SPURTS AND PHYSICAL DEVELOPMENT

One of the most visible signs of growth in preschoolers is the continued development of motor skills, both gross and fine. Gross motor skills, which involve larger muscle movements, become more refined as children gain strength, balance, and coordination. By the age of three, many children can run, jump, and climb with increased agility, and by age five, they can hop on one foot, skip, and even begin to ride a bicycle with training wheels. These activities not only indicate physical growth but also represent the increasing maturation of the nervous system, which allows children to better control their bodies and movements.

Fine motor skills, which involve smaller, more precise movements of the hands and fingers, also improve significantly during the preschool years. Children become more adept at tasks such as drawing, cutting with scissors, and manipulating small objects like beads or building blocks. By age four, many children can use utensils with greater dexterity, dress themselves with minimal assistance, and begin to write letters and numbers. These skills are essential for school readiness, as they lay the foundation for learning to write, draw, and engage in more complex tasks.

In addition to motor development, the preschool years are a time of significant growth in cognitive and language abilities. As children's brains continue to grow and form new connections, they become more capable of complex thinking and problem-solving. By age five, many children are able to follow multi-step instructions, categorize objects by various characteristics (such as size, shape, or color), and engage in imaginative play that reflects their understanding of the world around them. Language development also accelerates during this period, with children expanding their vocabulary to include several thousand words and becoming more adept at forming complete sentences, asking questions, and engaging in conversations.

The steady physical changes that occur during the preschool years are closely linked to cognitive and social development. As children gain greater control over their bodies and movements, they become more confident in their abilities and more eager to explore their surroundings. This increased independence is an important milestone, as it allows children to take on new challenges and responsibilities, both at home and in social settings like preschool or daycare.

Early Emotional Development

Emotional development is a critical aspect of growth during the preschool years, as children begin to understand and manage their feelings, develop empathy for others, and navigate social interactions. While every child develops at their own pace, the emotional milestones reached during this period are foundational for building healthy relationships and developing a positive sense of self.

One of the key emotional developments in preschoolers is the growing ability to regulate emotions. At age three, children may still have difficulty managing strong emotions like frustration, anger, or sadness, often leading to temper tantrums or outbursts. However, as they approach four and five years old, they typically become more capable of identifying their feelings and expressing them in more socially appropriate ways. For example, a five-year-old may still feel frustrated when things don't go their way, but they are more likely to use words to express their frustration or seek help from an adult, rather than resorting to a tantrum.

This period is also marked by the development of empathy, as preschoolers begin to understand that other people have feelings, thoughts, and perspectives that may differ from their own. Around the age of four, many children start to show concern for others when they are upset or hurt, offering comfort or trying to make amends when they recognize that their actions have caused distress. This ability to empathize is an important step toward building meaningful relationships and navigating social situations.

Preschoolers' increasing emotional awareness is often reflected in their play. Children at this age engage in more complex forms of pretend play, often taking on roles or characters that allow them to explore different emotions and scenarios. Through role-playing, they learn to navigate social dynamics, experiment with different ways of expressing themselves, and practice problem-solving in a safe and imaginative environment. This type of play is not only essential for emotional development but also for cognitive growth, as it encourages children to think creatively and critically.

Another key aspect of emotional development during the preschool years is the formation of a sense of identity. Children begin to see themselves as individuals with unique qualities, preferences, and abilities. They may express strong opinions about what they like and dislike, and they often take pride in their accomplishments, such as learning a new skill or completing a task independently. This growing sense of self contributes to the development of self-esteem, which is shaped by positive feedback from caregivers, teachers, and peers.

Despite the significant progress preschoolers make in regulating their emotions and understanding others, they still require guidance and support from adults to navigate difficult situations. Caregivers play a crucial role in modeling appropriate emotional responses, teaching problem-solving skills, and providing reassurance during challenging moments. By offering consistent support and encouragement, caregivers help preschoolers build the emotional resilience needed to face new challenges and develop healthy relationships.

Signs of Delayed Growth

While most children progress through the preschool years at their own unique pace, there are certain signs that may indicate delayed growth or development. Identifying these signs early is important, as it allows for timely intervention and support that can help children catch up to their peers and reach their full potential.

PRESCHOOL GROWTH (3-5 YEARS)

Delayed physical growth can manifest in several ways. For example, if a child is significantly shorter or lighter than other children their age, it may be a sign of a growth delay. While some variation in size is normal due to genetics, children who consistently fall below the 5th percentile on growth charts may require further evaluation. Other signs of delayed physical growth include failure to gain weight or height at the expected rate, a lack of energy, or persistent fatigue. In some cases, delayed growth may be caused by underlying medical conditions, such as hormonal imbalances, nutritional deficiencies, or chronic illnesses, that require medical attention.

In addition to physical growth, delays in motor skill development can also be a cause for concern. If a child struggles with tasks such as running, jumping, or climbing, or if they have difficulty with fine motor skills like holding a pencil or using scissors, it may indicate a delay in motor development. Delayed motor skills can impact a child's ability to participate in physical activities and may affect their confidence and social interactions.

Cognitive and language delays are another potential area of concern during the preschool years. If a child is not speaking in full sentences by age four or has difficulty understanding basic instructions, it may signal a delay in language development. Similarly, if a child struggles with basic problem-solving tasks or seems uninterested in playing with others or engaging in imaginative play, it may indicate a delay in cognitive or social development.

Emotional delays can also become apparent during the preschool years. Children who have difficulty regulating their emotions, experience frequent tantrums or meltdowns, or struggle to form relationships with peers may benefit from additional support. While occasional emotional outbursts are normal for preschoolers, persistent difficulties in managing emotions or engaging with others can be a sign of developmental delays that may require intervention.

Caregivers who notice any signs of delayed growth or development should consult with a pediatrician or a developmental specialist for further evaluation. Early intervention programs, such as speech therapy, occupational therapy, or physical therapy, can help children overcome delays and catch up to their peers. The earlier these interventions are implemented, the more effective they are likely to be in supporting the child's overall development.

Encouraging Healthy Eating and Activity

Fostering healthy eating habits and encouraging regular physical activity are essential for supporting growth and development during the preschool years. Proper nutrition provides the building blocks needed for physical growth, brain development, and overall health, while physical activity helps strengthen muscles, improve coordination, and promote a healthy weight.

Preschoolers have small stomachs and high energy needs, which means they require nutrient-dense foods that provide the vitamins, minerals, and calories necessary for growth. A balanced diet for preschoolers should include a variety of fruits, vegetables, whole grains, lean proteins, and healthy fats. Caregivers should aim to offer meals and snacks that are rich in nutrients and low in added sugars, unhealthy fats, and processed foods.

One of the challenges during the preschool years is managing picky eating. Many children become more selective about the foods they are willing to eat during this phase, often preferring familiar or plain foods over more adventurous options. While picky eating is a normal part of development, caregivers can encourage healthy eating by offering a variety of foods, modeling good eating habits, and involving children in meal preparation. Presenting foods in fun and creative ways, such as cutting vegetables into interesting shapes or arranging colorful plates, can also make meals more appealing to young children.

It is important to establish regular meal and snack times, as preschoolers benefit from a consistent eating schedule. Offering three meals and two healthy snacks each day helps ensure that children receive the nutrients they need to support their growth and development throughout the day. Caregivers should also encourage children to listen to their hunger and fullness cues, allowing them to eat when they are hungry and stop when they are full. This helps preschoolers develop a healthy relationship with food and avoid overeating, which can lead to weight issues later in life.

Hydration is another important aspect of healthy eating. Water should be the primary beverage offered to preschoolers, with milk being another good option to provide calcium and vitamin D. Sugary drinks, such as sodas and fruit juices, should be limited, as they contribute to excess calorie intake without providing essential nutrients. Instead, offering whole fruits and vegetables as snacks ensures that children receive the fiber and vitamins they need in a more nutritious form.

Physical activity is equally important for preschoolers, as it helps build strong muscles and bones, supports motor skill development, and promotes overall physical health. Preschoolers should have opportunities for both structured and unstructured physical play every day. Structured activities can include games like tag, playing ball, or participating in organized sports, while unstructured activities involve free play, such as running, climbing, or exploring the outdoors. Both types of play are essential for fostering physical development and encouraging children to enjoy movement.

The preschool years are a critical time to establish a foundation for lifelong physical activity. Caregivers can help by making physical activity a fun and enjoyable part of everyday life. Trips to the playground, dancing to music, and simple obstacle courses set up at home are just a few ways to get preschoolers moving. Additionally, involving children in family activities, such as walking or biking, sets a positive example of the importance of staying active.

Screen time should be limited during the preschool years to ensure that children have enough time for active play and physical movement. While some educational programming can be beneficial, excessive screen time can interfere with a child's ability to engage in physical activity and can contribute to sedentary behavior. The American Academy of Pediatrics recommends that screen time for children aged 2 to 5 be limited to one hour per day of high-quality programming, with caregivers actively engaging with the content to help children understand and apply what they are learning.

In addition to promoting physical health, physical activity during the preschool years supports cognitive and social development. Active play helps preschoolers practice problem-solving skills, improve concentration, and develop social skills through interactions with peers. Games that involve taking turns, following rules, and working together with others are especially beneficial for social-emotional development, as they teach children how to cooperate, negotiate, and resolve conflicts.

Ensuring that preschoolers have plenty of opportunities for both physical activity and rest is key to their overall growth and development. Preschoolers typically need 10 to 13 hours of sleep each night, including naps, to support their physical and cognitive development. A consistent bedtime routine and a calm, quiet environment can help preschoolers get the rest they need to grow and thrive.

In conclusion, encouraging healthy eating and regular physical activity during the preschool years helps support the steady physical and cognitive growth that takes place during this critical developmental period. By providing a balanced diet, promoting active play, and fostering healthy sleep habits, caregivers can help preschoolers develop the skills and habits they need for a lifetime of health and well-being.

In summary, the preschool years are a time of significant growth and development across multiple domains, including physical, cognitive, emotional, and social areas. While growth during this phase is more gradual than in the earlier years, it is nonetheless crucial for setting the foundation for future development. Caregivers play a key role in supporting preschoolers through this process, providing the guidance, nutrition, and encouragement needed to help children reach their full potential. By fostering a nurturing environment that prioritizes healthy habits, caregivers can ensure that preschoolers thrive during these formative years and are well-prepared for the next stages of childhood.

School-Age Growth Spurts (6-9 Years)

Between the ages of six and nine, children experience a steady progression in physical growth, mental development, and social maturation, which sets the stage for more significant changes during the pre-adolescent years. This phase, commonly referred to as middle childhood, serves as a critical period for establishing foundational habits in health, emotional regulation, and learning. While the rapid growth spurts of early childhood have slowed, children continue to grow at a gradual pace, and the importance of fostering physical activity, emotional well-being, and intellectual engagement becomes increasingly clear. This chapter explores the key aspects of growth for school-age children, the role of physical activity in supporting development, the shifts in mental and social behavior, and how parents can prepare for the onset of puberty in the later stages of childhood.

Gradual Physical Growth

As children enter middle childhood, their physical growth becomes more predictable and consistent, with most experiencing steady increases in height and weight each year. On average, children grow about 2 to 2.5 inches in height annually and gain around 4 to 7 pounds per year. This consistent, gradual growth helps children develop more proportional bodies, as the trunk, arms, and legs lengthen and become more balanced in relation to their head. By this stage, the chubbiness associated with early childhood fades, and children begin to take on a leaner, more athletic appearance.

SCHOOL-AGE GROWTH SPURTS (6-9 YEARS)

Although the pace of growth during these years is steady, individual rates can vary based on a combination of genetics, nutrition, and overall health. Some children may experience growth spurts earlier than others, while others may progress more slowly but catch up in later years. It's not uncommon for children to go through phases where they seem to grow more quickly in height, followed by a period of weight gain as their bodies adjust to new proportions. This fluctuation is normal and should not cause concern unless a child falls significantly below or above the average growth curve, at which point a pediatrician may recommend further evaluation.

The growth of bones and muscles during this phase is particularly important, as children build the skeletal and muscular framework that will support them into adolescence and adulthood. Calcium and vitamin D are essential nutrients during this time, as they play a key role in the development of strong bones. Ensuring that children get enough dairy products, leafy greens, and fortified foods helps support healthy bone growth. Additionally, exposure to sunlight aids in the natural production of vitamin D, which is crucial for calcium absorption.

Another critical aspect of physical growth in middle childhood is dental development. Between the ages of six and nine, most children lose their primary teeth (commonly known as baby teeth) and begin to grow permanent teeth. This process, while gradual, marks a significant milestone in physical development and can occasionally cause discomfort. Regular dental care, including brushing, flossing, and check-ups, becomes increasingly important during this time to ensure healthy oral development.

Despite the predictability of growth during middle childhood, certain factors can influence the pace of development. Genetics plays a primary role in determining a child's ultimate height and body composition, but environmental factors, such as nutrition, physical activity, and overall health, also play significant roles. Children who are malnourished or experience chronic illnesses may show slower growth, while those who have access to balanced nutrition and regular physical activity are more likely to reach their full growth potential. Regular monitoring of height, weight, and other growth markers during pediatric check-ups helps ensure that children are developing as expected.

Importance of Physical Activity

Physical activity is an essential component of healthy growth and development during middle childhood. As children grow stronger and more coordinated, they naturally become more active, and it's important to provide them with opportunities to engage in physical activities that promote muscle strengthening, cardiovascular fitness, and overall physical health. The habits children form during this time can have a lasting impact on their physical well-being, as regular physical activity sets the foundation for a healthy, active lifestyle into adolescence and adulthood.

During middle childhood, children typically develop more advanced gross motor skills, such as running, jumping, throwing, and catching. These activities require coordination, strength, and balance, which improve as children grow and mature. Participating in organized sports, such as soccer, basketball, or gymnastics, not only helps children build physical skills but also teaches important life lessons, such as teamwork, perseverance, and discipline. Additionally, unstructured play, such as riding bikes, climbing trees, or playing on playground equipment, allows children to explore their physical capabilities and engage in creative, active play.

Physical activity is not only important for building strong muscles and bones, but it also plays a critical role in maintaining a healthy weight. As children's metabolism slows slightly compared to their earlier years, they become more susceptible to weight gain if they do not balance caloric intake with physical activity. Regular exercise helps burn excess calories and prevents the accumulation of body fat, reducing the risk of childhood obesity, which has become a growing concern in many parts of the world. Establishing a routine that includes at least one hour of moderate to vigorous physical activity each day helps children maintain a healthy weight and develop good habits for the future.

In addition to its physical benefits, exercise has been shown to improve cognitive function and academic performance. Studies suggest that children who engage in regular physical activity demonstrate better concentration, memory, and problem-solving abilities. Physical activity increases blood flow to the brain, which promotes the growth of new brain cells and enhances neural connections, ultimately supporting learning and cognitive development. Schools that incorporate regular physical education and opportunities for active play throughout the day often see improved behavior and academic outcomes among students.

Another key benefit of physical activity is its impact on mental health. Children who participate in regular exercise are more likely to experience lower levels of stress, anxiety, and depression. Physical activity stimulates the production of endorphins, which are natural mood enhancers, and helps children manage their emotions in a healthy way. Additionally, participating in team sports or group activities fosters social connections, which can contribute to a child's sense of belonging and self-esteem.

Parents and caregivers play a crucial role in encouraging physical activity during middle childhood. Providing access to a variety of physical activities, both structured and unstructured, helps children explore their interests and develop their physical abilities. Encouraging outdoor play, limiting screen time, and participating in family activities like hiking or swimming can all help promote an active lifestyle. It's also important for parents to model positive behaviors by staying active themselves and making physical activity a regular part of family life.

Mental and Social Development

Mental and social development undergo significant transformations during middle childhood, as children begin to refine their cognitive abilities and navigate more complex social relationships. Between the ages of six and nine, children become more independent, capable of critical thinking, and socially aware, which prepares them for the challenges of adolescence.

One of the key developments during this phase is the shift from concrete to more abstract thinking. While younger children tend to focus on the here and now, school-age children begin to understand concepts that are not immediately observable, such as time, space, and cause and effect. This cognitive shift allows children to engage in more complex problem-solving and logical reasoning. For example, a child might be able to think through the steps needed to complete a task or consider different possible outcomes of a situation before taking action. This ability to think ahead and plan is an important skill that will continue to develop into adolescence.

Children's memory and attention span also improve during this phase, allowing them to absorb and retain more information. This is especially important for academic success, as children in this age group are expected to master basic skills in reading, writing, and mathematics. As their cognitive abilities expand, children become more curious about the world around them, often asking deeper questions and seeking to understand how things work. This curiosity is a key driver of learning, and caregivers can support it by providing opportunities for exploration, such as reading books, visiting museums, or engaging in hands-on activities that encourage problem-solving and creativity.

Social development during middle childhood is marked by an increasing awareness of others and the formation of more complex peer relationships. Children at this age begin to move beyond the egocentric thinking of early childhood and start to consider the perspectives, thoughts, and feelings of others. This growing empathy allows them to build stronger friendships and navigate social situations with greater ease. Peer relationships become more important during this time, and children often form close-knit groups with friends who share similar interests.

While friendships become more central to a child's social life, school-age children also begin to develop a deeper understanding of social norms and rules. They learn about fairness, cooperation, and the importance of following rules in both structured environments, such as school or sports teams, and in unstructured play with peers. These social interactions help children practice important skills, such as communication, negotiation, and conflict resolution, which are essential for building healthy relationships.

Despite their growing independence, children in middle childhood still rely on guidance from adults to navigate social and emotional challenges. Parents, teachers, and caregivers play a crucial role in helping children develop positive self-esteem, emotional regulation, and problem-solving skills. Open communication, encouragement, and support help children build confidence and resilience as they face new challenges, both academically and socially.

It's also important to recognize that some children may struggle with social development during this time. Shyness, difficulty making friends, or problems with social cues may indicate that a child needs additional support. Encouraging participation in group activities, providing opportunities for social interaction, and teaching social skills through role-playing or modeling can help children build the confidence they need to engage with their peers.

Preparing for Puberty: What Parents Should Know

As children near the end of middle childhood, the first signs of puberty may begin to emerge, especially for girls. While the full onset of puberty typically occurs between the ages of nine and 12, some children may begin to experience early signs of hormonal changes as young as eight or nine. This phase, often referred to as pre-puberty, marks the beginning of significant physical and emotional changes that will continue into adolescence.

For girls, the earliest signs of puberty may include the development of breast buds and the growth of pubic hair. These changes usually occur around age eight or nine, although they can happen earlier or later depending on the individual. Girls may also experience a slight increase in body fat, particularly around the hips and thighs, as their bodies prepare for menstruation. While these changes are normal, they can sometimes cause anxiety or confusion for young girls, especially if they occur earlier than expected. Parents should be prepared to discuss these changes in a supportive and age-appropriate manner, reassuring their children that these are natural stages of development.

For boys, the onset of puberty generally begins a little later than for girls, typically around the ages of nine to 11. The first noticeable sign of puberty in boys is the enlargement of the testicles, followed by the growth of pubic hair. Boys may also begin to experience a growth spurt, with increased muscle mass and height, though these changes typically occur more dramatically in the later stages of puberty. During this pre-pubescent phase, boys may also experience an increase in body odor and the early development of acne, both of which are triggered by hormonal changes. Like girls, boys may feel self-conscious about these changes, especially if they occur before their peers. Open communication with parents about the physical changes of puberty can help alleviate concerns and provide reassurance.

Both boys and girls may experience emotional changes during this pre-puberty phase as well. The hormonal fluctuations that accompany early puberty can lead to mood swings, irritability, and increased sensitivity. Children may become more self-conscious about their appearance or more concerned about fitting in with their peers. These emotional shifts are a normal part of development, but they can sometimes cause tension between children and their parents or peers. It's important for parents to recognize these emotional changes and offer patience, understanding, and support as their children navigate the transition into adolescence.

SCHOOL-AGE GROWTH SPURTS (6-9 YEARS)

One of the most important things parents can do to prepare their children for puberty is to begin open and honest conversations about the physical and emotional changes they can expect. By discussing these topics early and in a supportive manner, parents can help normalize the experience of puberty and reduce the anxiety children may feel about the changes in their bodies. Parents should aim to create a safe space where children feel comfortable asking questions and expressing any concerns they may have about puberty.

In addition to discussing the physical changes of puberty, parents should also address topics such as hygiene, emotional regulation, and social pressures. As children's bodies begin to change, they will need to adopt new hygiene habits, such as more frequent bathing, using deodorant, and caring for their skin to prevent acne. Parents can also help children develop strategies for managing the emotional ups and downs that may accompany hormonal changes, such as practicing mindfulness, engaging in physical activity, or talking through their feelings with a trusted adult.

Finally, it's important for parents to monitor their children's physical and emotional health as they approach puberty. While most children will experience a smooth transition into adolescence, some may encounter challenges, such as early or delayed puberty, which can impact their self-esteem and social interactions. If parents notice that their child is struggling with the changes of puberty, or if puberty seems to be occurring much earlier or later than expected, it may be helpful to consult with a pediatrician or a healthcare provider. Early or delayed puberty can sometimes indicate underlying health issues that require medical attention, and addressing these concerns early can help support the child's overall well-being.

In conclusion, the ages of six to nine mark a critical period of development where steady physical growth, mental advancement, and social skills come into sharper focus. During these years, children build the foundation for healthy physical habits, emotional regulation, and social interactions, all of which will serve them well as they transition into adolescence. By understanding the importance of gradual growth, encouraging physical activity, fostering mental and social development, and preparing for the onset of puberty, parents and caregivers can help children navigate this phase with confidence and resilience.

Puberty and Its Physical Impact (13-15 Years

Between the ages of 10 and 12, children undergo significant physical, emotional, and cognitive changes that prepare them for the transition into adolescence. This period, often referred to as pre-adolescence or the pre-teen years, is marked by the onset of puberty, which brings about noticeable growth spurts, changes in body proportions, hormonal shifts, and heightened emotional development. While these changes occur at varying rates depending on the individual, this stage of development is crucial for setting the foundation for adolescence and beyond. Understanding the various aspects of pre-adolescent growth, including the role of nutrition and the impact of hormonal changes, helps caregivers support children through this complex and trans-formative phase.

Pre-Puberty Signs and Growth Spurts

One of the most defining characteristics of pre-adolescence is the onset of puberty, which typically begins between the ages of 8 and 13 for girls and 9 and 14 for boys. Puberty marks the beginning of sexual maturation, driven by hormonal changes that lead to the development of secondary sexual characteristics, growth spurts, and other physiological transformations. While the exact timing of puberty varies among individuals due to genetic, environmental, and nutritional factors, most children begin to experience the early signs of puberty during the pre-adolescent years.

GROWTH SPURTS AND PHYSICAL DEVELOPMENT

The most noticeable sign of pre-puberty is the growth spurt, a period of rapid physical growth in both height and weight. For many children, this growth spurt occurs between the ages of 10 and 12, although it may begin earlier for girls and slightly later for boys. During this time, children may grow several inches in height over the course of a year or two, often resulting in noticeable changes in their appearance. Boys and girls typically experience their growth spurts at different times, with girls often beginning their growth spurt earlier, around ages 10 to 11, while boys tend to experience theirs closer to 12.

This accelerated growth is driven by the increased production of growth hormones, including human growth hormone (HGH) and insulin-like growth factor (IGF-1), both of which play critical roles in bone and muscle development. These hormones stimulate the lengthening of bones, particularly the long bones in the arms and legs, which contributes to the dramatic increases in height seen during this stage. The growth spurt also involves changes in muscle mass, as the body begins to prepare for the more pronounced physical changes of adolescence.

While the growth spurt is a natural part of development, it can sometimes lead to feelings of awkwardness or discomfort for pre-teens. Children may experience growing pains, characterized by aches or discomfort in the legs or arms, as their bones and muscles stretch and adapt to rapid growth. These growing pains are generally temporary and can often be relieved with rest, gentle stretching, or over-the-counter pain relievers.

Additionally, the uneven pace of growth can lead to temporary imbalances in coordination. As children grow taller, their limbs may lengthen faster than their bodies, leading to a period of adjustment where they may appear clumsy or awkward. This is particularly common in boys, who may experience rapid growth in height before their muscles catch up in strength and coordination. While this imbalance is typically short-lived, it can affect a child's self-esteem during a time when they are becoming increasingly aware of their bodies and physical abilities.

Another important sign of pre-puberty is the development of secondary sexual characteristics, which vary between boys and girls. For girls, these early signs often include the development of breast buds, the appearance of pubic hair, and changes in body fat distribution, particularly around the hips and thighs. Boys, on the other hand, may experience the enlargement of the testicles and penis, the growth of pubic hair, and an increase in muscle mass. Both boys and girls may also experience changes in skin texture and oil production, which can lead to the onset of acne. While these physical changes are normal and expected, they can sometimes cause anxiety or embarrassment, especially if a child feels that they are developing faster or slower than their peers.

Changes in Body Proportions

As children enter the pre-adolescent years, one of the most noticeable changes is the alteration of body proportions. Prior to puberty, children's bodies are relatively proportional, with their heads, trunks, and limbs growing at similar rates. However, as they approach adolescence, their bodies begin to change in preparation for the more significant physical developments that occur during puberty.

For girls, the most prominent changes in body proportions often involve an increase in body fat, particularly around the hips, thighs, and buttocks. This redistribution of fat is part of the body's preparation for eventual childbearing, and it results in a more defined waistline and curvier figure. The growth of breast tissue is another key change in body proportions during this time, with the development of breast buds marking the beginning of breast growth that will continue through adolescence. While these changes are normal and necessary for female development, they can sometimes lead to body image concerns, particularly if a girl feels self-conscious about her changing shape. It is important for caregivers to provide reassurance and promote body positivity during this time, helping girls understand that these changes are a natural part of growing up.

For boys, changes in body proportions during pre-adolescence are typically less pronounced than in girls, as boys generally begin their physical maturation slightly later. However, boys may still experience some early changes in body composition, including a slight increase in muscle mass and the lengthening of their limbs. As boys approach puberty, their shoulders may begin to broaden, and their overall body shape will shift to a more masculine form. Boys also experience an increase in body fat during this period, but it is generally less noticeable than the changes seen in girls. Like girls, boys may feel self-conscious about their bodies during this stage, especially if they feel that they are not developing as quickly as their peers.

Another important aspect of changing body proportions during pre-adolescence is the growth of the hands and feet. It is not uncommon for pre-teens to experience a rapid increase in the size of their hands and feet before the rest of their body catches up. This can lead to feelings of awkwardness or self-consciousness, as children may feel that their bodies are out of proportion. However, these changes are typically temporary, and the rest of the body will eventually grow to match the larger hands and feet. Caregivers can help children navigate these changes by reassuring them that their bodies are developing normally and that these feelings of awkwardness are a common part of growing up.

As children's bodies change, it is important to recognize that these shifts in body proportions are not just physical—they also have psychological and social implications. Pre-teens are becoming increasingly aware of their appearance and how they compare to others, and they may begin to develop concerns about fitting in with their peers. Body image issues can emerge during this time, particularly as children are exposed to societal ideals of beauty and physical appearance through media and social interactions. Caregivers and educators play a critical role in promoting a healthy body image and helping children understand that their worth is not tied to their appearance. Encouraging open conversations about body changes and emphasizing the importance of self-acceptance can help pre-teens build a positive relationship with their bodies as they continue to grow and develop.

Hormonal Shifts and Emotional Development

The physical changes that occur during pre-adolescence are driven by hormonal shifts that also have a significant impact on emotional development. As children approach puberty, their bodies begin to produce higher levels of sex hormones, including estrogen in girls and testosterone in boys. These hormones play a key role in the development of secondary sexual characteristics and reproductive organs, but they also influence mood, behavior, and emotional regulation.

One of the most notable effects of these hormonal changes is the increase in mood swings and emotional sensitivity. Pre-teens may experience intense emotions, such as irritability, frustration, or sadness, which can sometimes seem out of proportion to the situation. These mood swings are often unpredictable and may leave children feeling confused or overwhelmed by their own emotions. Additionally, hormonal fluctuations can affect a child's ability to concentrate and manage stress, which can impact their performance at school and their relationships with peers and family members.

While these emotional changes are a normal part of development, they can sometimes lead to conflicts at home or in school. Pre-teens may become more argumentative, defiant, or withdrawn as they navigate the complexities of their emotions. It is important for caregivers to recognize that these behaviors are often a reflection of the child's struggle to cope with new feelings and experiences. Providing a supportive and understanding environment, where children feel safe to express their emotions, can help them navigate this challenging time.

In addition to mood swings, pre-teens may begin to experience feelings of self-consciousness and insecurity, particularly as their bodies change and they become more aware of how they are perceived by others. Peer relationships become increasingly important during this time, and children may worry about fitting in, being accepted, or meeting societal expectations. This heightened sensitivity to social dynamics can lead to anxiety or feelings of inadequacy, especially if a child feels that they are not developing at the same rate as their peers.

Caregivers can help support emotional development during pre-adolescence by encouraging open communication and helping children develop healthy coping strategies. Teaching pre-teens how to manage stress, regulate their emotions, and build resilience can empower them to navigate the challenges of growing up. Additionally, fostering a sense of self-worth that is based on more than physical appearance can help pre-teens develop a positive self-image and avoid the negative effects of peer pressure or societal expectations.

The Role of Nutrition During Pre-Adolescence

Nutrition plays a crucial role in supporting the physical and emotional changes that occur during pre-adolescence. As children enter their growth spurts and experience hormonal changes, their bodies require additional nutrients to fuel healthy development. Ensuring that pre-teens receive a balanced diet that includes essential vitamins, minerals, and macro nutrients is key to supporting their growth and overall well-being.

One of the most important nutrients during pre-adolescence is calcium, which plays a critical role in the development of strong bones. As children undergo growth spurts, particularly in their limbs, the demand for calcium increases significantly. This mineral is essential for ensuring that bones grow properly and maintain their density. During pre-adolescence, children are in a key window for building bone mass, which can protect against future bone health issues like osteoporosis. Dairy products, such as milk, cheese, and yogurt, are excellent sources of calcium, but other options like leafy green vegetables, fortified cereals, and plant-based milks also provide this essential nutrient.

Along with calcium, vitamin D is equally important as it aids in calcium absorption. Without adequate vitamin D, the body struggles to utilize the calcium from food sources, which can hinder bone development. While some vitamin D can be obtained through diet, sunlight exposure is one of the primary sources of this vitamin. Encouraging outdoor play and ensuring that children have some exposure to sunlight (while taking appropriate skin protection measures) helps support their vitamin D levels. Foods such as fatty fish, eggs, and fortified products can also help increase vitamin D intake.

Protein is another vital component of a pre-adolescent's diet. It is responsible for the growth and repair of tissues, including muscles, and is essential for supporting the increased muscle mass that occurs during this phase. Protein-rich foods like lean meats, poultry, fish, beans, legumes, eggs, and nuts should be included in daily meals to ensure that children get the building blocks they need for healthy growth. During growth spurts, a child's body uses protein to build not only muscles but also to support the development of organs and other tissues. For boys, especially, who often begin to develop more muscle mass during this phase, adequate protein intake is particularly important.

Iron is another nutrient that becomes more important as children near puberty, especially for girls who will soon begin menstruation. Iron is needed for the production of hemoglobin, which carries oxygen in the blood. As the body grows and muscle mass increases, the demand for iron rises. Insufficient iron intake can lead to anemia, a condition that causes fatigue, weakness, and reduced concentration, which can affect school performance and overall well-being. Foods like red meat, poultry, fish, lentils, beans, and fortified cereals are good sources of iron. Including vitamin C-rich foods, like oranges or strawberries, in meals can enhance iron absorption from plant-based sources.

Carbohydrates, the body's primary energy source, are also important for pre-teens, particularly as their physical activity levels increase. However, it is crucial to emphasize complex carbohydrates from whole grains, fruits, and vegetables rather than refined sugars and processed foods. Complex carbohydrates provide a steady release of energy and are packed with fiber, vitamins, and minerals, all of which are important for sustaining energy throughout the day, especially during school hours or physical activities. Fiber-rich foods also promote digestive health, which is essential as children's bodies process more nutrients during their growth phases.

Fats, particularly healthy fats such as those found in olive oil, avocados, nuts, and fatty fish, are also an important part of a balanced diet. These fats support brain development, which continues throughout adolescence, and help regulate hormones. Omega-3 fatty acids, found in fish like salmon and walnuts, are especially beneficial for brain health and cognitive function. Limiting unhealthy fats, such as trans fats and excessive saturated fats found in processed foods, helps protect against future health problems like cardiovascular disease.

While ensuring that pre-adolescents consume a balanced and nutrient-dense diet is important, it is equally crucial to teach them about healthy eating habits. As children grow more independent, they begin making more choices about what they eat, both at school and at home. This is a key time for establishing lifelong healthy eating behaviors. Encouraging children to make balanced choices, understand portion sizes, and listen to their hunger and fullness cues can help prevent overeating and promote a healthy relationship with food.

Parents and caregivers can foster these habits by involving children in meal planning and preparation, which gives them a sense of ownership over their food choices and helps them understand the importance of nutrition. Providing a variety of nutritious options at home, packing balanced school lunches, and modeling healthy eating behaviors are all ways to encourage pre-teens to develop healthy habits. It's also important to approach food and nutrition with positivity, avoiding food shaming or overly restrictive diets, which can lead to disordered eating patterns later in life.

During pre-adolescence, it is not uncommon for children to develop food preferences or become more selective about what they eat. While picky eating can be frustrating, it's important to approach it with patience and creativity. Offering a variety of foods, introducing new items in fun and interesting ways, and encouraging children to try new things without pressuring them can help expand their palates. Providing positive reinforcement and making mealtime enjoyable rather than stressful fosters a healthy attitude toward food.

In addition to nutrition, hydration is a key component of supporting pre-adolescent growth and development. As children become more active and their bodies grow, their need for water increases. Encouraging regular water consumption throughout the day, particularly before and after physical activity, helps prevent dehydration and supports overall health. Limiting sugary drinks, such as sodas and sweetened juices, is also important to avoid unnecessary calories and protect dental health.

Ultimately, nutrition during pre-adolescence serves as the fuel for both physical growth and cognitive development. A balanced diet that provides all the essential nutrients is key to ensuring that children reach their full potential as they transition into adolescence. Supporting healthy eating habits, teaching the importance of balanced meals, and encouraging children to make mindful food choices will set the stage for a lifetime of good health.

In conclusion, the pre-adolescent years, typically between the ages of 10 and 12, are a time of profound physical, emotional, and cognitive change. As children approach puberty, they experience rapid growth spurts, changes in body proportions, and hormonal shifts that affect both their physical appearance and emotional well-being. Understanding the role of nutrition in supporting these changes is crucial for helping children navigate this transitional period successfully. By providing a supportive environment, encouraging open communication, and fostering healthy habits, parents and caregivers can help pre-adolescents thrive during this pivotal stage of development.

Late Adolescence (16-18 Years)

Adolescence marks one of the most trans-formative phases in human development, encompassing dramatic physical, emotional, and cognitive changes as individuals transition from childhood into adulthood. Between the ages of 13 and 15, young people experience puberty, a biological process driven by hormonal changes that trigger growth spurts, sexual maturation, and a range of other physical and emotional shifts. These years are particularly significant because they not only shape the physical body but also deeply influence an individual's sense of self, relationships, and behavior. The rapid changes that occur during this stage can be overwhelming, both for adolescents and for their caregivers, and understanding the nature of these developments is crucial for supporting healthy growth and emotional well-being.

Major Growth Spurts in Height and Weight

One of the most noticeable aspects of puberty is the dramatic growth spurt that occurs, particularly in height and weight. For most adolescents, these growth spurts represent some of the most rapid periods of physical development since infancy. The timing of growth spurts varies, with girls typically beginning their major growth around ages 10 to 12 and boys starting later, around ages 12 to 14. By the time they reach 15, most adolescents are nearing their final adult height, although some may continue growing into their late teens.

LATE ADOLESCENCE (16-18 YEARS)

During this phase, boys and girls experience increases in height at different rates, with girls often completing their growth spurts by age 16 and boys continuing to grow until age 18 or beyond. Boys can grow as much as 4 to 5 inches per year during their peak growth spurt, while girls tend to grow more slowly, typically gaining about 2 to 3 inches annually. The growth plates in the long bones, especially in the arms and legs, remain open during this period, allowing for the dramatic increase in height that is characteristic of adolescence. The process is regulated by growth hormones, including human growth hormone (HGH) and sex hormones such as estrogen and testosterone.

In addition to increases in height, adolescents also experience changes in body composition, particularly weight gain. For girls, puberty is often associated with an increase in body fat, particularly around the hips, thighs, and breasts, as the body prepares for reproductive maturity. This shift in body fat distribution contributes to the development of a more defined waist and curvier figure. For boys, the growth spurt is accompanied by an increase in muscle mass, particularly in the upper body. Testosterone, the primary male sex hormone, stimulates the development of larger muscles and a broader chest and shoulders. This increase in muscle mass often contributes to a more angular and muscular appearance.

Although these changes are a normal part of development, they can sometimes be a source of discomfort or insecurity for adolescents. Rapid weight gain or height increases can lead to feelings of awkwardness, as young people may struggle to adjust to their changing bodies. Boys, in particular, may experience periods of clumsiness as their muscles and coordination take time to catch up with their increased height. Girls may feel self-conscious about the redistribution of body fat and the onset of menstruation, both of which are highly visible signs of physical maturation.

The variability in the timing of growth spurts can also lead to comparisons among peers, which can further heighten self-consciousness. Adolescents who develop earlier or later than their peers may feel isolated or different, leading to social or emotional challenges. For example, boys who are shorter or less physically developed than their classmates may feel insecure about their appearance, while girls who experience early breast development may feel uncomfortable with the attention they receive. It is important for caregivers and educators to offer reassurance during this time and emphasize that physical development occurs at different rates for everyone.

Sexual Maturity and Physical Changes

In addition to growth spurts, puberty involves significant sexual maturation, driven by the increased production of sex hormones. These hormones—testosterone in boys and estrogen in girls—trigger the development of secondary sexual characteristics, including changes in the reproductive organs and the development of visible signs of sexual maturity.

For girls, the onset of menstruation, or menarche, is one of the most significant milestones of puberty. Most girls experience their first menstrual period between the ages of 10 and 15, with the average age being around 12 or 13. Menstruation is a clear indication that the body is becoming capable of reproduction, as it signifies the beginning of ovulation. In addition to the start of menstruation, girls also experience breast development, which typically begins a few years earlier. The development of breast tissue occurs in stages, starting with the appearance of breast buds and continuing until full breast maturation is achieved, usually by age 16 or 17.

Along with breast development and menstruation, girls also experience the growth of pubic and underarm hair, which is stimulated by the increase in androgens, a group of hormones that includes testosterone. Body fat increases during this time, especially around the hips and thighs, which contributes to the development of a more feminine body shape. The reproductive organs, including the ovaries, fallopian tubes, and uterus, also mature during this time, although these changes are not as visibly apparent as external developments.

LATE ADOLESCENCE (16-18 YEARS)

For boys, puberty is marked by the enlargement of the testicles and penis, which typically begins around age 11 or 12. This growth is one of the first signs of sexual maturity, followed by the development of pubic, underarm, and facial hair. Testosterone, the primary male sex hormone, is responsible for these changes, as well as the deepening of the voice, which occurs as the vocal cords lengthen and the larynx grows larger. Boys also experience an increase in muscle mass and strength during puberty, driven by the surge in testosterone. The development of secondary sexual characteristics, such as facial and body hair, can continue into the late teens and early twenties, with some boys not achieving full sexual maturity until their early twenties.

Erections and nocturnal emissions, commonly known as "wet dreams," are another aspect of male sexual development during puberty. These involuntary events are a normal part of growing up and are often a source of confusion or embarrassment for boys. Open communication with caregivers about sexual health and development can help normalize these experiences and reduce anxiety around sexual maturation.

For both boys and girls, the development of sexual characteristics and the ability to reproduce bring about new responsibilities and questions about sexual health. Adolescents may begin to experience sexual attraction and curiosity, making it important for caregivers to provide accurate and age-appropriate information about sexual health, consent, and healthy relationships. Fostering an environment of open dialogue about sexual development can help adolescents make informed choices as they navigate this new stage of life.

Acne, Voice Changes, and Body Odor: Managing Puberty Challenges

The physical changes of puberty are not limited to growth spurts and sexual development; adolescents also experience a range of other changes that can impact their daily lives and self-esteem. Acne, body odor, and voice changes are common challenges during puberty, all of which are related to the increased production of hormones.

Acne is one of the most common skin conditions affecting adolescents, with around 85% of teens experiencing some form of acne during puberty. The condition is caused by an increase in androgens, which stimulate the sebaceous glands to produce more oil (sebum). When excess oil combines with dead skin cells, it can clog pores, leading to the development of pimples, blackheads, and cysts. Acne often appears on the face, back, and chest, and its severity can range from mild to severe. While acne is a normal part of puberty, it can significantly affect an adolescent's self-esteem, especially if they feel self-conscious about their appearance.

Managing acne during puberty involves proper skin care and, in some cases, medical treatment. Washing the face with a gentle cleanser, avoiding harsh scrubbing, and using non-comedogenic (non-pore-clogging) products can help reduce the severity of acne. For more severe cases, over-the-counter treatments containing benzoyl peroxide or salicylic acid may be effective. If acne persists or causes significant distress, a dermatologist can recommend prescription treatments, such as topical retinoids or oral medications. It's important for adolescents to understand that acne is a temporary condition and that their skin will likely improve as they age.

In addition to acne, body odor is another common challenge during puberty. As the body produces more sweat due to increased activity in the sweat glands, bacteria on the skin break down the sweat, leading to odor. This is particularly noticeable in areas like the underarms, feet, and groin. Adolescents may become more self-conscious about body odor during puberty, especially as they become more socially aware and concerned about their appearance.

Managing body odor involves practicing good hygiene habits, such as showering regularly, using deodorant or antiperspirant, and wearing clean clothes. Encouraging adolescents to develop a daily hygiene routine can help them feel more confident and comfortable during this time of physical change. In some cases, excessive sweating (hyperhidrosis) may require medical intervention, but for most teens, regular hygiene practices are sufficient for managing body odor.

For boys, one of the most noticeable changes during puberty is the deepening of the voice. This occurs as the larynx, or voice box, grows larger and the vocal cords lengthen in response to the increase in testosterone. The process is often gradual, with boys experiencing voice "cracks" or fluctuations in pitch as their vocal cords adjust. This can be a source of embarrassment for some boys, especially if the changes occur during social interactions or public speaking. However, the voice typically stabilizes by the end of puberty, resulting in a deeper, more mature sound.

While these physical changes are normal, they can contribute to feelings of awkwardness or insecurity during adolescence. Offering reassurance and normalizing the experience of puberty can help alleviate some of the anxiety associated with these changes. Caregivers and educators can also provide practical advice for managing the physical challenges of puberty, such as using appropriate skin care products, developing a hygiene routine, and understanding that voice changes are temporary.

Emotional Roller coaster: Understanding Adolescent Mood Swings

The physical changes of puberty are accompanied by a range of emotional and psychological shifts that can make adolescence feel like an emotional roller coaster for many teenagers. These mood swings are largely driven by the hormonal fluctuations that occur during puberty, with increased levels of estrogen and testosterone affecting the brain's neurotransmitter systems, particularly those involved in regulating mood, stress, and emotions. Adolescents often experience heightened emotions, which can manifest as irritability, frustration, sadness, or euphoria, sometimes shifting rapidly and unpredictably. These emotional changes are not only a result of hormonal activity but are also influenced by the social and cognitive developments that occur during this stage of life.

One of the most significant emotional challenges during adolescence is the increased sensitivity to social dynamics and peer relationships. As adolescents develop a stronger sense of identity, they also become more concerned with how they are perceived by others, particularly their peers. Friendships and social status take on greater importance, and adolescents may experience intense feelings of belonging or exclusion, depending on their social interactions. This heightened sensitivity can lead to emotional turbulence, as adolescents navigate the complexities of friendships, romantic interests, and group dynamics.

Social comparison is a common feature of adolescence, as teenagers often compare themselves to their peers in terms of physical appearance, academic achievements, athletic abilities, and social standing. These comparisons can sometimes lead to feelings of inadequacy or low self-esteem, particularly if an adolescent feels they do not measure up to societal or peer expectations. This is especially true for body image, as the physical changes of puberty can lead to self-consciousness about weight, height, skin appearance, or other aspects of the body. For girls, the pressure to conform to certain beauty standards can be especially intense, while boys may feel pressure to develop a muscular physique or achieve physical milestones at the same rate as their peers.

In addition to peer pressure, adolescents also begin to experience more complex emotions related to identity and self-concept. During this period, teenagers are in the process of forming their own identities, often questioning previously held beliefs and values as they explore different aspects of their personality. This process of self-discovery can sometimes lead to internal conflicts, as adolescents struggle to reconcile their emerging sense of self with societal or parental expectations. The desire for independence and autonomy often clashes with the need for parental guidance and support, leading to tension in family relationships. Adolescents may become more argumentative or defiant as they assert their independence, which can be

LATE ADOLESCENCE (16-18 YEARS)

challenging for caregivers who are adjusting to their child's changing needs.

Despite the emotional ups and downs that characterize adolescence, this period of heightened emotions also presents opportunities for growth. Adolescents develop greater emotional intelligence during this time, learning to identify and express their feelings, manage stress, and navigate interpersonal relationships. Emotional experiences during adolescence are important for developing resilience and coping skills, which will serve individuals throughout their lives. Caregivers can support emotional development by providing a safe and non-judgmental environment where teenagers feel comfortable expressing their emotions. Encouraging open communication and active listening helps adolescents feel understood and supported during this tumultuous time.

It's important for caregivers to recognize that while mood swings and emotional sensitivity are common during adolescence, persistent or extreme emotional difficulties may require additional support. Mental health issues, such as depression, anxiety, or eating disorders, often emerge during adolescence, and early intervention is key to addressing these challenges. If an adolescent appears to be struggling with prolonged feelings of sadness, hopelessness, or other emotional difficulties, it may be helpful to consult a mental health professional. Ensuring that adolescents have access to emotional support, whether from family, friends, or counselors, is critical for helping them navigate the emotional challenges of this stage.

Another aspect of emotional development during adolescence is the increased capacity for empathy and deeper emotional connections with others. As teenagers develop more abstract thinking and cognitive abilities, they begin to understand others' perspectives in more nuanced ways, which enhances their ability to form close relationships. Romantic relationships often become more significant during this time, and adolescents may experience the intense emotions that come with first love, infatuation, or heartbreak. These experiences, while emotionally charged, are important for developing relational skills and learning how to navigate complex emotions in a social context.

While the emotional roller coaster of adolescence can be challenging for both teenagers and their caregivers, it is also a period of tremendous growth. Adolescents are learning to manage new emotions, assert their independence, and form their identities, all of which are crucial steps in their journey toward adulthood. By offering guidance, support, and understanding, caregivers can help teenagers navigate the emotional highs and lows of puberty with confidence and resilience.

In conclusion, the years between 13 and 15 are marked by profound physical and emotional changes as adolescents experience puberty and the accompanying growth spurts, sexual maturation, and emotional shifts. These years can be challenging, as teenagers grapple with body image concerns, social pressures, and intense mood swings. However, this period also offers opportunities for growth, as adolescents learn to manage their emotions, form deeper connections with others, and develop a stronger sense of self. By providing support and fostering open communication, caregivers can help adolescents navigate this pivotal stage of development, setting the foundation for healthy emotional and physical well-being as they move toward adulthood.

Nutrition for Optimal Growth

Late adolescence, the period between the ages of 16 and 18, is a time of significant growth, both physically and emotionally, as teenagers transition from adolescence to adulthood. This phase represents the final stages of physical maturation, including the completion of growth spurts and the solidification of muscle mass and bone strength. It is also a period marked by increasing emotional and mental maturity as adolescents develop a more stable sense of self, refine their ability to manage emotions, and gain independence. However, challenges such as body image issues, social pressures, and the need for guidance as they prepare for adulthood remain key aspects of this stage. Understanding how late adolescence shapes both short- and long-term health outcomes is crucial for supporting individuals during this pivotal time.

Final Growth Spurts: Height, Muscle Development, and Bone Strength

By the age of 16, most girls have reached their final adult height, with the majority of their height growth having occurred during earlier adolescent growth spurts. For boys, however, the growth process may continue until around 18 years old or even into their early twenties in some cases. Boys often experience their peak growth spurt a bit later than girls, which can result in noticeable changes in their height and body composition during this period. The closing of the growth plates in the bones, known as epiphyseal fusion, signifies the end of vertical growth for both sexes.

During late adolescence, muscle development becomes more prominent, particularly in boys. Testosterone, which plays a key role in male puberty, continues to promote muscle mass development, resulting in increased upper body strength and overall physical maturation. While girls also continue to build muscle, their muscle growth tends to be less pronounced due to lower levels of testosterone. However, both sexes benefit from physical activity, such as sports or strength training, during this time, as these activities help to build and maintain muscle tone.

Bone strength is another critical aspect of physical development during late adolescence. The accumulation of bone density during this period is crucial for long-term bone health, as individuals build up to 90% of their peak bone mass by the end of adolescence. Adequate calcium intake, along with physical activity, particularly weight-bearing exercises like walking, running, or lifting weights, is essential for promoting strong bones. Vitamin D, which helps the body absorb calcium, is also important, and regular exposure to sunlight or supplementation can support healthy bone development.

While most adolescents complete their growth in height by the end of this period, their bodies continue to develop in other ways, particularly in terms of muscle tone and the distribution of body fat. For boys, increased testosterone levels help build lean muscle mass and reduce body fat, while girls may see a further accumulation of fat in areas such as the hips, thighs, and breasts, which are associated with reproductive health. These changes in body composition are a natural part of development, but they can also influence how adolescents perceive their bodies and their self-esteem.

Understanding Body Image Issues

Body image issues are particularly prevalent during late adolescence, as individuals become more aware of societal standards of beauty and often compare themselves to their peers or media representations of the "ideal" body. The physical changes that accompany the end of puberty, such as shifts in weight distribution, muscle tone, and the final stages of height growth, can lead to feelings of self-consciousness or dissatisfaction with one's appearance. For many adolescents, the pressure to conform to certain body ideals can be overwhelming and contribute to negative body image, low self-esteem, and even the development of eating disorders or unhealthy behaviors.

Girls, in particular, may face societal pressures to maintain a slim figure, even as their bodies naturally increase in fat as part of reproductive maturation. The expectation to fit a specific mold, coupled with unrealistic portrayals of women's bodies in the media, can result in dissatisfaction with weight or shape, even if an individual is within a healthy range. This dissatisfaction may manifest in behaviors such as excessive dieting, disordered eating, or over-exercising as adolescents strive to meet unattainable standards of beauty. It's important to note that body image concerns are not limited to girls—boys also experience pressures related to physical appearance, particularly in regard to muscle mass and leanness.

For boys, the pressure to build muscle and appear physically strong can lead to an overemphasis on achieving a muscular physique. Boys may feel inadequate if they do not meet societal expectations of strength or athleticism, leading to anxiety about their body image. The rise of social media, where edited and filtered images of "ideal" bodies are often highlighted, exacerbates these feelings of inadequacy for both boys and girls. In extreme cases, body dissatisfaction in boys can lead to behaviors such as the use of performance-enhancing drugs or supplements to increase muscle mass, which can have serious long-term health consequences.

Addressing body image issues during late adolescence requires a multifaceted approach that includes promoting body positivity, encouraging healthy lifestyle habits, and fostering self-acceptance. It is essential to teach adolescents that bodies come in all shapes and sizes and that health is not defined solely by appearance. Helping teenagers understand that the images they see in the media are often manipulated can also reduce the pressure to conform to unrealistic standards. Additionally, caregivers and educators can encourage physical activity not as a means to change one's appearance, but as a way to support overall health and well-being.

Mental health support is also critical for adolescents struggling with body image issues. Open communication about body changes and feelings of self-worth, coupled with counseling or therapy if necessary, can provide valuable tools for managing these challenges. By creating an environment that emphasizes self-acceptance and healthy habits, caregivers can help mitigate the effects of societal pressure on body image during this vulnerable time.

Mental and Emotional Maturity

While late adolescence is characterized by continued physical development, it is also a time of profound mental and emotional growth. By the ages of 16 to 18, adolescents are beginning to develop a stronger sense of identity, independence, and personal responsibility. They become more adept at critical thinking, problem-solving, and understanding abstract concepts, such as ethics and justice. Cognitive abilities such as planning, organizing, and decision-making improve as the prefrontal cortex, the part of the brain responsible for executive functions, continues to mature.

Emotionally, adolescents in this age group are better able to regulate their emotions and respond to stress more effectively than they could during early adolescence. While mood swings and emotional sensitivity are still common due to the ongoing hormonal changes of late adolescence, most teenagers begin to develop greater emotional stability as they learn to cope with their feelings. This emotional maturity allows them to form deeper and more meaningful relationships, both with peers and with family members.

One of the key aspects of emotional development during late adolescence is the process of identity formation. Adolescents in this stage often spend a significant amount of time exploring different aspects of their identity, such as their values, beliefs, interests, and social roles. This exploration is a critical part of the process of becoming an independent adult. Some teenagers may experiment with different social groups, hobbies, or even appearance as they try to figure out who they are and where they fit in the world. This period of self-discovery is often accompanied by a search for autonomy, as adolescents seek to differentiate themselves from their parents and establish their own path.

While this quest for independence is a normal part of development, it can also lead to conflicts with caregivers as adolescents push against boundaries and challenge authority. The desire to make independent decisions about aspects of their lives, such as education, friendships, and career aspirations, may clash with parental expectations or guidance. It is important for caregivers to strike a balance between providing support and allowing adolescents the freedom to make their own choices, even if those choices involve making mistakes.

In addition to their growing independence, adolescents are also developing a more nuanced understanding of social relationships. By this age, most teenagers are capable of forming intimate relationships, both romantic and platonic, that are based on mutual respect and emotional connection. However, navigating these relationships can be challenging, as adolescents are still learning how to balance their own needs with the needs of others. Friendships may become more stable and emotionally supportive during this time, but they can also be a source of stress if conflicts arise.

Romantic relationships, in particular, can be a significant aspect of late adolescence. While these relationships provide opportunities for emotional growth and intimacy, they can also introduce new challenges, such as dealing with rejection, navigating sexual boundaries, or managing the complexities of love and affection. Caregivers can play an important role in guiding adolescents through these experiences by fostering open communication about healthy relationships, consent, and emotional well-being.

Mental health is a crucial consideration during late adolescence, as this is a time when many mental health issues, such as depression, anxiety, and substance use, may emerge or intensify. The pressure to succeed academically, fit in socially, or meet personal goals can contribute to feelings of stress and overwhelm. It is important for caregivers and educators to be aware of the signs of mental health challenges and to provide appropriate support when needed. Encouraging healthy coping mechanisms, such as mindfulness, physical activity, or seeking help from a trusted adult, can help adolescents manage stress and build emotional resilience.

Preparing for Adulthood: How Growth Affects Long-Term Health

As adolescents approach the end of their teenage years, they are on the cusp of adulthood and are beginning to establish habits and behaviors that will influence their long-term health and well-being. The physical and emotional growth that occurs during late adolescence plays a critical role in shaping an individual's future, particularly in terms of health outcomes and lifestyle choices.

One of the most important aspects of late adolescence is the opportunity to solidify healthy habits, such as regular physical activity, balanced nutrition, and adequate sleep. Adolescents who engage in these behaviors are more likely to carry them into adulthood, reducing their risk of chronic health conditions such as obesity, heart disease, and diabetes. Conversely, unhealthy habits established during this time, such as poor diet, lack of exercise, or substance use, can have long-lasting negative effects on health.

Bone health is one area where late adolescence has a particularly significant impact on long-term health. As mentioned earlier, late adolescence is the final phase of bone development, during which individuals reach peak bone mass. This is a critical period for building strong bones, which will help protect against osteoporosis and fractures later in life. Adolescents who do not get enough calcium, vitamin D, or engage in weight-bearing physical activities may not reach their full bone density potential, which could make them more vulnerable to bone health issues as they age. Establishing habits such as consuming calcium-rich foods like dairy products, leafy greens, and fortified foods, as well as engaging in regular exercise, helps set the stage for a lifetime of healthy bones.

In addition to bone health, maintaining a healthy body weight during adolescence is important for preventing future health problems. Adolescents who struggle with weight issues, whether underweight or overweight, are at increased risk for a variety of health conditions, including cardiovascular disease, diabetes, and metabolic syndrome. Late adolescence is an ideal time to help young people develop a healthy relationship with food and exercise, emphasizing balance and moderation rather than restrictive diets or unhealthy weight-loss practices. Encouraging adolescents to view food as fuel for their growing bodies and to understand the importance of staying active can promote long-term physical health.

Sleep is another key factor that influences long-term health. Although adolescents are notorious for having irregular sleep patterns due to school schedules, social activities, and changing biological rhythms, getting enough sleep is essential for both physical and mental health. During late adolescence, the body continues to grow and repair itself during sleep, and insufficient rest can negatively impact academic performance, mood, and overall well-being. Chronic sleep deprivation has been linked to a higher risk of obesity, diabetes, heart disease, and mental health disorders later in life. Helping adolescents establish good sleep hygiene, such as maintaining a consistent bedtime, limiting screen time before bed, and creating a relaxing sleep environment, is essential for their immediate and future health.

Mental health is equally important as physical health in late adolescence, and emotional well-being during this time can have long-lasting effects. Adolescents who develop healthy coping mechanisms for dealing with stress, anxiety, and emotional challenges are more likely to maintain good mental health as they transition into adulthood. On the other hand, individuals who do not have access to emotional support or who engage in harmful coping strategies, such as substance use or self-isolation, may struggle with mental health issues that persist into adulthood. It is crucial for caregivers, educators, and mental health professionals to be proactive in addressing mental health concerns, providing resources and support when necessary.

As adolescents prepare to transition into adulthood, they also begin to take more responsibility for their health care. This may include managing medical appointments, understanding their own health needs, and making decisions about preventive care. Helping adolescents learn to advocate for their own health and wellness is an important part of their development during this stage. Encouraging them to take ownership of their health, whether it's by scheduling regular check-ups, discussing vaccination options, or being aware of their family health history, helps prepare them for the responsibilities that come with adulthood.

Late adolescence is also a critical time for discussing and promoting sexual health. As adolescents become more sexually active, it is important that they have access to accurate information about contraception, sexually transmitted infections (STIs), and consent. Providing education on these topics helps adolescents make informed choices and protects their long-term reproductive and sexual health. Open, non-judgmental conversations about sexual health can reduce the risk of unintended pregnancies, STIs, and other negative health outcomes associated with unsafe sexual practices.

The development of emotional maturity during late adolescence also influences long-term mental and relational health. As teenagers solidify their identity and learn to manage their emotions, they lay the foundation for healthy interpersonal relationships in adulthood. Adolescents who learn to communicate effectively, practice empathy, and navigate conflicts in a constructive manner are more likely to build successful relationships with peers, romantic partners, and future colleagues. Moreover, individuals who develop a strong sense of self-worth and self-acceptance during late adolescence are better equipped to handle the challenges and transitions of adult life with resilience.

In addition to personal relationships, late adolescence is often a time when individuals begin to think seriously about their future career paths and life goals. Developing a sense of purpose and direction is an important aspect of emotional and cognitive maturity, and adolescents who are encouraged to explore their interests and talents are more likely to pursue careers that align with their passions and values. The decisions made during this period—whether related to education, employment, or personal goals—can have lasting implications for an individual's financial stability, personal fulfillment, and overall quality of life.

Preparing adolescents for adulthood involves not only supporting their physical and mental health but also equipping them with the skills and knowledge they need to succeed in the world. This includes practical life skills, such as managing finances, understanding nutrition, and practicing time management, as well as emotional and social skills, such as conflict resolution, self-advocacy, and maintaining healthy boundaries. Adolescents who are well-prepared for adulthood are more likely to thrive in their personal and professional lives and to make positive contributions to their communities.

In conclusion, late adolescence is a time of both physical and emotional transformation as individuals complete their growth spurts, develop muscle and bone strength, and begin to solidify their sense of identity and purpose. While this period is often accompanied by challenges related to body image, social pressures, and emotional volatility, it also provides opportunities for growth and self-discovery. By promoting healthy habits, supporting emotional development, and helping adolescents prepare for the responsibilities of adulthood, caregivers and educators can guide young people toward a future of physical health, emotional well-being, and personal success.

Genetics and Growth

Growth is influenced by a combination of genetic, environmental, and hormonal factors, with genetics playing a particularly significant role in determining an individual's potential for height, body composition, and even the timing of developmental milestones. Understanding how hereditary influences shape growth can help explain why individuals of the same age may vary significantly in size, body shape, and the rate at which they mature. While most variations in growth are a normal part of human diversity, certain genetic conditions can affect normal growth patterns, leading to the need for medical intervention or specialist care. This chapter explores the role of genetics in growth, highlights genetic conditions that impact growth, and provides guidance on when it might be necessary to consult a specialist.

Understanding Hereditary Influences

Growth is largely predetermined by an individual's genetic blueprint, with approximately 60-80% of a person's height and overall growth pattern being inherited from their parents. Genes regulate the production of growth hormones, the sensitivity of tissues to these hormones, and the development of bones and muscles. While environmental factors such as nutrition, physical activity, and overall health also play an important role, a person's genetic makeup establishes the upper and lower boundaries of their growth potential.

One of the most straightforward examples of genetic influence on growth is seen in familial patterns of height. Taller parents tend to have taller children, and shorter parents tend to have shorter children. This tendency is due to the inheritance of multiple genes that regulate bone growth, the timing of growth spurts, and the cessation of growth after puberty. Children inherit these traits from both parents, which explains why siblings—despite having the same parents—can exhibit variations in height, depending on which combination of growth-related genes they inherit.

In addition to height, genetics can also influence body composition, such as the distribution of fat and muscle, as well as metabolic rate. Some individuals may inherit a predisposition for a leaner build with more muscle mass, while others may be genetically inclined to store more fat. These genetic differences often become more noticeable during puberty, when the body undergoes significant changes in response to hormonal signals that are also influenced by genetic factors. For example, boys who inherit higher levels of testosterone tend to build more muscle mass, while girls who inherit genes associated with higher estrogen levels may develop more body fat, particularly around the hips and thighs.

Another critical aspect of hereditary influence on growth is the timing of puberty. The onset of puberty, which marks a period of rapid growth, is also largely controlled by genetics. Early or late puberty can run in families, with individuals often following similar patterns to their parents or close relatives. This means that children whose parents experienced early puberty may also go through puberty earlier than their peers, while those whose parents developed later may not experience the same growth spurts until their late teens. These variations in pubertal timing can impact overall growth and the final adult height, as early bloomers may reach their full height sooner, while late bloomers may continue to grow for a longer period.

Although genetics provides the blueprint for growth, it is important to remember that it does not act in isolation. Environmental factors, such as diet, access to healthcare, and physical activity, can either enhance or inhibit genetic growth potential. For instance, children who are genetically predisposed to be tall may not reach their full height potential if they suffer from malnutrition or chronic illness during critical periods of growth. Conversely, children with average height potential may surpass expectations if they receive optimal nutrition and healthcare.

Genetic Conditions Affecting Growth

While genetic inheritance typically follows a predictable pattern that results in normal growth variation, certain genetic conditions can interfere with normal growth processes, leading to either excessive or stunted growth. These genetic conditions can affect the production of growth hormones, the development of bones and tissues, or the regulation of puberty and other growth-related processes. Below are some of the most common genetic conditions that impact growth:

1. Turner Syndrome:
Turner syndrome is a genetic disorder that affects only females and is caused by the complete or partial absence of one of the two X chromosomes. This condition occurs in approximately 1 in 2,500 live female births. Girls with Turner syndrome typically have short stature due to impaired growth hormone production and lack the typical pubertal growth spurt. In addition to short stature, they may experience delayed or absent puberty and have characteristic physical features such as a broad chest and webbed neck. Growth hormone therapy can often help increase height in girls with Turner syndrome if started early enough, and hormone replacement therapy may be needed to induce puberty.

2. Achondroplasia:
Achondroplasia is the most common form of dwarfism, affecting about 1 in 15,000 to 40,000 live births. It is caused by a mutation in the FGFR3 gene, which regulates bone growth. Individuals with achondroplasia have disproportionately short arms and legs, while their torso remains relatively

normal in size. Achondroplasia primarily affects the growth of the long bones, leading to short stature and a characteristic skeletal appearance. Despite their shorter stature, individuals with achondroplasia can lead healthy and fulfilling lives, although they may require specialized medical care for associated health issues such as spinal stenosis or joint problems.

3. Marfan Syndrome:

Marfan syndrome is a genetic disorder that affects the body's connective tissue, which provides structure and support to organs and tissues. Individuals with Marfan syndrome tend to be unusually tall and have long arms, legs, and fingers. The condition is caused by a mutation in the FBN1 gene, which affects the production of fibrillin, a protein that helps maintain the elasticity of connective tissue. In addition to excessive height, people with Marfan syndrome are at risk for serious health complications, including heart and blood vessel problems, due to the weakened connective tissue in the aorta and other major blood vessels. Early diagnosis and management are crucial for preventing life-threatening complications.

4. Klinefelter Syndrome:

Klinefelter syndrome is a condition that affects males and occurs when an individual has one or more extra X chromosomes (XXY instead of the typical XY). This condition affects approximately 1 in 500 to 1,000 live male births. Males with Klinefelter syndrome tend to be taller than average but may have reduced muscle mass, broader hips, and delayed puberty. They may also have learning difficulties or mild cognitive impairments. Hormone replacement therapy, particularly testosterone, is often used to support puberty and muscle development in individuals with Klinefelter syndrome.

5. Noonan Syndrome:

Noonan syndrome is a genetic disorder that affects multiple parts of the body and can result in short stature, heart defects, and distinct facial features such as a wide-set eyes, a small jaw, and low-set ears. It affects approximately 1 in 1,000 to 2,500 live births. Growth hormone therapy can help individuals with Noonan syndrome reach a more typical height for their age, but early diagnosis is essential for effective treatment.

6. Sotos Syndrome:

Sotos syndrome, also known as cerebral gigantism, is a rare genetic condition characterized by excessive growth during childhood, leading to children who are significantly taller than their peers. The condition is caused by mutations in the NSD1 gene, which affects growth and development. In addition to overgrowth, children with Sotos syndrome may experience developmental delays, learning disabilities, and distinct facial features. While growth may normalize in adulthood, individuals with Sotos syndrome often require ongoing support for developmental challenges.

7. Growth Hormone Deficiency (GHD):

Growth hormone deficiency is a condition in which the pituitary gland does not produce enough growth hormone, resulting in stunted growth and short stature. GHD can be genetic or acquired later in childhood due to injury or illness. Children with GHD grow at a much slower rate than their peers and may have immature facial features. Growth hormone therapy is the primary treatment for GHD and can significantly improve height outcomes if started early.

When to Consult a Specialist

While variations in growth are often normal and influenced by genetic diversity, there are certain signs and symptoms that may indicate the need for specialist care. Early identification of growth-related issues can make a significant difference in treatment outcomes, particularly when it comes to conditions that are responsive to hormone therapy or other interventions.

Parents and caregivers should consider consulting a pediatric endocrinologist or genetic specialist if they notice any of the following signs:

1. Delayed or Accelerated Growth:

If a child is significantly shorter or taller than their peers, or if their growth appears to have stalled or accelerated at an unusual rate, it may be a sign of an underlying genetic or hormonal condition. Growth charts, which track a child's height and weight over time, can help identify whether a child is following a normal growth pattern. If a child consistently falls below the 5th percentile or above the 95th percentile for their age, it may warrant further investigation.

2. Delayed Puberty:

While the timing of puberty varies, most girls begin puberty by age 13 and most boys by age 14. If a child has not shown any signs of puberty by these ages, it may indicate a hormonal or genetic issue that requires evaluation. Delayed puberty can be caused by conditions like Turner syndrome, Klinefelter syndrome, or growth hormone deficiency.

3. Excessive or Abnormal Growth Patterns:

Excessive growth or abnormal body proportions may indicate conditions like Marfan syndrome, Sotos syndrome, or other disorders that affect bone development and connective tissue. If a child grows significantly taller than expected for their age, has disproportionately long limbs, or exhibits other unusual growth patterns, a genetic evaluation may be necessary.

4. Unusual Physical Features or Developmental Delays:

Certain genetic conditions that affect growth may also cause distinct physical features or developmental delays. For example, children with Noonan syndrome or achondroplasia often have characteristic facial features or skeletal abnormalities. Developmental delays, such as difficulties with motor skills, speech, or learning, may also signal a genetic condition that impacts growth and overall development. If a child exhibits any of these signs alongside abnormal growth patterns, it is essential to consult with a specialist who can assess the situation and determine whether genetic testing or other evaluations are needed.

5. Familial History of Growth Disorders:

   A family history of growth-related disorders, such as Turner syndrome, Marfan syndrome, or achondroplasia, may increase the likelihood of a child inheriting similar conditions. If parents or close relatives have experienced growth disorders or genetic conditions that affect development, consulting a specialist early can help in monitoring the child's growth and providing timely interventions if necessary.

6. Growth Hormone Deficiency Symptoms:

   Children with growth hormone deficiency may show signs of slowed growth, delayed tooth eruption, or immature facial features. These children typically do not experience the normal growth spurts expected during childhood and adolescence. If a child exhibits these symptoms, a pediatric endocrinologist may perform tests to evaluate growth hormone levels and recommend hormone therapy if needed.

The Diagnostic Process

When a child is referred to a specialist due to concerns about their growth, the diagnostic process typically begins with a thorough medical history and physical examination. The specialist will review the child's growth charts, taking into account their height, weight, and rate of growth over time. Family history is also crucial, as many genetic conditions that affect growth are hereditary.

If a growth disorder is suspected, the next step often involves diagnostic tests such as blood tests to measure hormone levels (including growth hormone, thyroid hormones, and sex hormones) or genetic tests to identify specific chromosomal abnormalities. Imaging tests, such as X-rays, may be used to assess bone age, which can help determine whether a child's bones are developing normally relative to their chronological age. In some cases, more advanced imaging techniques, such as MRIs, may be required to examine the pituitary gland or other parts of the brain that regulate growth.

Once a diagnosis is made, the specialist will work with the family to develop a treatment plan. For many growth-related disorders, early intervention is key to achieving the best possible outcomes. For example, growth hormone therapy can significantly improve height outcomes for children with growth hormone deficiency or Turner syndrome, while hormone replacement therapy may be necessary for children with delayed or absent puberty. In cases of genetic conditions like achondroplasia or Marfan syndrome, ongoing medical monitoring and management may be required to address associated health issues.

Importance of Early Intervention

Early diagnosis and treatment of genetic growth disorders can have a profound impact on a child's quality of life. By addressing growth issues early, healthcare providers can help ensure that children reach their full potential in terms of height, physical development, and cognitive abilities. This is particularly true for conditions that are responsive to hormone therapies, such as growth hormone deficiency or Turner syndrome, where starting treatment during early childhood or adolescence can lead to better long-term outcomes.

Beyond physical growth, early intervention is also essential for addressing any developmental delays or health complications associated with genetic conditions. For example, children with Marfan syndrome are at risk for serious cardiovascular problems, and early monitoring and treatment can prevent life-threatening complications. Similarly, children with conditions like Noonan syndrome or Sotos syndrome may require additional support for developmental delays, and early access to educational and therapeutic resources can significantly improve their cognitive and social development.

For families, early intervention also provides the opportunity to better understand their child's condition and access the resources they need to support their child's growth and development. Genetic counseling may be recommended to help parents understand the inheritance patterns of certain conditions and the likelihood of recurrence in future pregnancies. Ongoing support from specialists, such as pediatric endocrinologists, geneticists, and developmental pediatricians, ensures that children with growth disorders receive comprehensive care tailored to their specific needs.

In summary, genetics plays a critical role in shaping growth, from determining height and body composition to influencing the timing of puberty and the onset of growth spurts. While most growth variations are within the range of normal human diversity, certain genetic conditions can significantly impact growth and development, requiring specialist care and intervention. Early diagnosis and treatment are essential for optimizing growth outcomes and addressing any associated health issues. By understanding the hereditary influences on growth and recognizing the signs of potential growth disorders, families and healthcare providers can work together to ensure that children receive the care and support they need to thrive.

Sleep and Growth

Sleep plays a critical role in the growth and development of children and adolescents. During sleep, the body undergoes essential processes that support physical growth, cognitive function, and emotional well-being. From the release of growth hormones to the restoration of muscles and tissues, sleep is deeply intertwined with a child's ability to grow and thrive. However, not all children get the necessary amount or quality of sleep to support optimal growth, and sleep disorders or poor sleep habits can have significant negative impacts on physical development. This chapter will explore the importance of sleep for growth, identify sleep disorders that may hinder development, and provide strategies for creating healthy sleep habits to ensure children are getting the rest they need to support their growth.

The Importance of Sleep for Physical Development

Sleep is essential for all aspects of health, but its role in physical development is particularly pronounced during periods of growth, such as infancy, childhood, and adolescence. One of the key reasons sleep is so important for growth is its role in the release of growth hormones. The majority of growth hormone production occurs during deep sleep, especially during the non-REM stages of the sleep cycle. Growth hormone is responsible for the development of muscles, bones, and tissues, and it plays a crucial role in overall physical maturation.

1. Growth Hormone Release During Sleep:

GROWTH SPURTS AND PHYSICAL DEVELOPMENT

Growth hormone is produced and released by the pituitary gland, a small gland located at the base of the brain. During deep sleep, particularly in the early part of the night, the pituitary gland releases pulses of growth hormone into the bloodstream. This hormone stimulates the growth of bones and tissues, helps to increase muscle mass, and supports the repair and regeneration of cells. For children and adolescents, who are in critical periods of growth, this process is essential for reaching their full height potential and developing strong bones and muscles. Inadequate sleep can disrupt the release of growth hormone, leading to slower growth and, in some cases, stunted development.

The impact of growth hormone on development is most evident during childhood and adolescence, when the body is growing rapidly. During these periods, children and teenagers require more sleep than adults to support the increased demands of their developing bodies. The American Academy of Sleep Medicine recommends that school-aged children (6-12 years old) get 9-12 hours of sleep per night, while teenagers (13-18 years old) need 8-10 hours per night. When children and adolescents do not get enough sleep, the amount of growth hormone released is reduced, which can negatively affect their physical development over time.

2. Cellular Repair and Immune Function:

Sleep is also crucial for the repair and restoration of cells and tissues. During sleep, the body works to repair damaged tissues, regenerate cells, and strengthen the immune system. For growing children and adolescents, this process is vital, as their bodies are constantly undergoing changes and experiencing physical demands from activities like sports, school, and daily play. Inadequate sleep can hinder the body's ability to repair itself, which may lead to slower recovery from injuries or illnesses and decreased overall physical resilience.

Furthermore, sleep plays a significant role in supporting the immune system. Research has shown that sleep helps the body produce cytokines, proteins that regulate immune responses. Cytokines are important for fighting infections and inflammation, and they are produced in greater quantities during sleep. For growing children, who are more susceptible to illnesses as their immune systems develop, getting enough sleep is essential for maintaining a strong and healthy immune response.

3. Bone Growth and Density:

Adequate sleep is particularly important for bone growth and density during childhood and adolescence. Bones grow and remodel throughout childhood, and this process is influenced by the release of growth hormone during sleep. During adolescence, when rapid growth spurts occur, the bones elongate as the growth plates (areas of developing tissue at the ends of long bones) expand. If a child or teenager consistently gets insufficient sleep, the growth plates may not receive adequate stimulation from growth hormone, potentially leading to delayed or stunted growth.

In addition to bone lengthening, sleep is crucial for the development of bone density. During sleep, the body absorbs and utilizes nutrients such as calcium and phosphorus, which are essential for building strong bones. This process helps children and adolescents build the bone mass that will protect them against conditions like osteoporosis later in life. A chronic lack of sleep can impair the body's ability to build strong bones, leading to a higher risk of fractures and bone health issues in the future.

4. Cognitive and Emotional Development:

While this chapter focuses primarily on the physical aspects of growth, it's important to acknowledge that sleep also plays a key role in cognitive and emotional development. For children and adolescents, who are learning new skills, acquiring knowledge, and navigating social relationships, sleep is essential for consolidating memories, improving attention, and regulating emotions. Inadequate sleep can lead to cognitive impairments, difficulty concentrating in school, mood swings, and increased stress, all of which can have indirect effects on physical growth and overall well-being.

Sleep Disorders That Affect Growth

Several sleep disorders can disrupt the normal sleep patterns of children and adolescents, leading to insufficient or poor-quality sleep. When sleep is consistently disrupted, it can hinder the release of growth hormones and impair physical development. Identifying and addressing these sleep disorders early is crucial for ensuring that children get the rest they need to support their growth.

1. Obstructive Sleep Apnea (OSA):

Obstructive sleep apnea is a common sleep disorder in which a person's airway becomes partially or completely blocked during sleep, causing breathing to stop and start repeatedly throughout the night. OSA can occur in both children and adults, but it is particularly concerning in growing children because it can significantly disrupt sleep quality and the release of growth hormone. Children with OSA may experience restless sleep, loud snoring, gasping for air during the night, and daytime sleepiness.

The frequent interruptions in breathing caused by OSA prevent the child from reaching deep, restorative stages of sleep, where the majority of growth hormone is released. As a result, children with untreated sleep apnea may experience slower growth, behavioral issues, and difficulties with concentration and learning. OSA is often associated with enlarged tonsils or adenoids, and in some cases, surgery to remove the tonsils and adenoids can resolve the condition. For children who are overweight, weight loss may also help alleviate symptoms of sleep apnea.

2. Restless Legs Syndrome (RLS):

Restless legs syndrome is a neurological disorder characterized by an irresistible urge to move the legs, usually in response to uncomfortable sensations. These sensations often occur in the evening or at night, making it difficult for children to fall asleep or stay asleep. Although RLS is more commonly associated with adults, it can also affect children and adolescents, particularly those with a family history of the disorder.

RLS can interfere with sleep quality, leading to insufficient deep sleep and reduced growth hormone production. Children with RLS may also experience fatigue, irritability, and difficulty concentrating during the day. Treating underlying causes of RLS, such as iron deficiency, can help improve symptoms, while establishing healthy sleep routines may reduce the impact of the disorder on sleep quality.

3. Insomnia:

Insomnia is characterized by difficulty falling asleep, staying asleep, or waking up too early and being unable to return to sleep. Insomnia can occur in children and adolescents due to a variety of factors, including stress, anxiety, and irregular sleep schedules. Chronic insomnia can lead to sleep deprivation, which negatively impacts growth by reducing the amount of time spent in deep sleep, where growth hormone release occurs.

Children and adolescents with insomnia may experience slower growth, irritability, difficulty concentrating, and a weakened immune system. Behavioral interventions, such as cognitive-behavioral therapy for insomnia (CBT-I), can be effective in helping children and teenagers develop better sleep habits and address the underlying causes of insomnia.

4. Delayed Sleep Phase Syndrome (DSPS):

Delayed sleep phase syndrome is a circadian rhythm disorder in which a person's internal body clock is shifted later than the typical sleep-wake cycle. Adolescents are particularly prone to DSPS, as their natural sleep-wake cycles tend to shift later during puberty. However, when this delayed sleep phase leads to chronic sleep deprivation—due to school schedules or other obligations—it can negatively affect growth and development.

Teenagers with DSPS may struggle to fall asleep at a reasonable time, resulting in insufficient sleep if they are required to wake up early for school. Over time, this lack of sleep can impair growth hormone release and hinder physical development. Treatment for DSPS often involves gradually shifting the sleep-wake cycle earlier, using light therapy, and practicing good sleep hygiene to support a more regular sleep schedule.

Creating Healthy Sleep Habits for Growing Kids

Establishing healthy sleep habits from an early age is essential for supporting physical growth, cognitive development, and overall well-being. While each child's sleep needs may vary slightly, creating a consistent routine that prioritizes sufficient sleep can help children and adolescents get the rest they need to grow and thrive. Here are some key strategies for promoting healthy sleep habits:

1. Create a Consistent Sleep Schedule:
Consistency is key when it comes to promoting healthy sleep. Encouraging children to go to bed and wake up at the same time every day helps regulate their internal body clock and supports more restful sleep. This consistency should be maintained even on weekends and holidays to prevent disruptions to the sleep-wake cycle.

2. Establish a Relaxing Bedtime Routine:
A relaxing bedtime routine can signal to the body that it's time to wind down and prepare for sleep. Activities such as reading a book, taking a warm bath, or listening to calming music can help children transition from the activities of the day to a restful night's sleep. Avoiding stimulating activities, such as watching TV or playing video games, in the hour before bed can also support better sleep.

3. Limit Screen Time Before bedtime:
Exposure to screens—whether from smartphones, tablets, computers, or televisions—can interfere with the body's production of melatonin, the hormone responsible for regulating sleep. The blue light emitted from these

devices can delay the onset of sleep by confusing the body's internal clock, making it harder for children and adolescents to fall asleep at the appropriate time. To promote healthy sleep, it's important to limit screen time in the hour or two leading up to bedtime. Encouraging children to engage in non-screen activities, such as reading or listening to calming music, can help them relax and prepare for sleep.

4. Create an Optimal Sleep Environment:

The bedroom environment plays a crucial role in promoting restful sleep. The ideal sleep environment is cool, quiet, dark, and comfortable. Parents can help by ensuring that their child's room is conducive to sleep by lowering the temperature, using blackout curtains to block out light, and minimizing noise. In some cases, a white noise machine or a fan can help drown out disruptive noises, creating a more peaceful atmosphere. Ensuring that the bed is comfortable and free from distractions (such as toys or devices) can also encourage better sleep.

5. Encourage Physical Activity During the Day:

Regular physical activity is important not only for physical health but also for promoting good sleep. Engaging in physical activities such as running, playing sports, or simply spending time outdoors can help children expend energy, making it easier for them to fall asleep at night. However, it's important to avoid vigorous physical activity close to bedtime, as it can have a stimulating effect and make it harder to wind down. Instead, aim for physical activity earlier in the day to support better sleep.

6. Monitor and Limit Caffeine Intake:

Caffeine is a stimulant that can interfere with sleep by making it harder to fall asleep and reducing the overall quality of sleep. Many children and teenagers consume caffeine in the form of soda, energy drinks, coffee, or tea, often without realizing the impact it can have on their sleep. To promote healthy sleep habits, it's important to limit caffeine intake, especially in the afternoon and evening. Encouraging children to opt for water or caffeine-free alternatives can help minimize the effects of caffeine on their sleep.

7. Teach Relaxation Techniques:

For children and adolescents who experience anxiety or stress that interferes with their sleep, teaching relaxation techniques can be an effective way to promote calm and prepare the mind and body for rest. Techniques such as deep breathing, progressive muscle relaxation, or guided imagery can help reduce feelings of stress and anxiety, making it easier to fall asleep. Practicing these techniques as part of a nightly routine can help children learn to manage their emotions and improve the quality of their sleep.

8. Promote a Balanced Sleep-Wake Cycle:

For teenagers, who naturally experience a shift in their circadian rhythm during puberty, it's important to encourage good sleep hygiene practices that promote a balanced sleep-wake cycle. Teenagers may be tempted to stay up late and sleep in on weekends, but this can disrupt their internal body clock and make it harder to maintain a consistent sleep schedule. While some flexibility is normal, encouraging teenagers to stick to a regular sleep schedule most days of the week can help them get the sleep they need for optimal growth and development.

The Role of Parents and Caregivers in Supporting Healthy Sleep

Parents and caregivers play a vital role in helping children and adolescents develop healthy sleep habits. By setting a good example and creating a home environment that prioritizes sleep, parents can instill the importance of rest and encourage lifelong healthy habits. Here are some ways that parents can support their child's sleep:

- Model Good Sleep Habits: Children learn by observing their parents, so setting a good example by maintaining your own consistent sleep schedule can reinforce the importance of sleep. Demonstrating healthy sleep practices, such as limiting screen time before bed and creating a relaxing bedtime routine, shows children how to prioritize rest.

- Educate Children About Sleep: Teaching children about the importance of sleep for their physical and mental health can help them understand why sleep should be a priority. Explaining how sleep affects their growth, ability to learn, and mood can encourage children to take ownership of their sleep habits.

- Address Sleep Issues Early: If a child is experiencing difficulties with sleep, such as frequent night wakings, insomnia, or symptoms of sleep apnea, it's important to address these issues early. Consulting with a pediatrician or sleep specialist can help identify the underlying causes of sleep problems and provide appropriate treatment or interventions.

In conclusion, sleep is a fundamental component of physical growth and development, playing a critical role in the release of growth hormones, cellular repair, immune function, and bone health. Sleep disorders, such as obstructive sleep apnea and insomnia, can disrupt these processes, potentially hindering growth and overall well-being. By establishing healthy sleep habits early on and addressing any sleep issues as they arise, parents and caregivers can help children and adolescents get the rest they need to support their growth and thrive throughout their developmental years.

Signs of Growth Delay

Growth is a dynamic and multifaceted process that varies from one individual to another. While children typically follow a predictable pattern of growth, deviations from this norm can sometimes indicate underlying health issues. Understanding what constitutes normal growth, how to recognize early signs of growth delay, and the methods used to diagnose growth disorders is crucial for ensuring timely interventions. This chapter explores the definition of normal growth, the early signs of growth disorders, and the diagnostic process for identifying delayed growth in children and adolescents.

What Is Considered Normal Growth?

Normal growth refers to the process by which a child increases in height, weight, and overall body size in a way that follows predictable patterns based on age, sex, and genetics. Children grow at varying rates throughout their lives, with periods of rapid growth—such as during infancy and puberty—followed by more gradual growth in other stages. Growth patterns are influenced by a range of factors, including genetics, nutrition, hormones, and environmental conditions.

1. Growth Charts and Percentiles:
    Pediatricians use growth charts to track a child's growth over time, comparing their height, weight, and body mass index (BMI) to standardized norms. These growth charts are based on data collected from large populations of

children and are separated by age and sex to account for normal variations in growth between boys and girls. Each child's measurements are plotted on the growth chart, which provides percentile rankings. For example, a child in the 50th percentile for height is taller than 50% of their peers and shorter than the other 50%.

Children typically follow a consistent growth trajectory along their own percentile range. For example, a child who has consistently measured in the 25th percentile for height is likely to remain in that range throughout childhood and adolescence. Fluctuations within this range are normal, but significant deviations—such as a child who drops from the 50th percentile to the 10th percentile in height—may indicate a growth problem. Maintaining a steady growth pattern, even if it's on the lower or higher end of the percentile range, is typically a sign of healthy development.

2. Expected Growth Rates by Age:
   Growth occurs in different phases throughout childhood, with distinct patterns at each stage of development:

- Infancy (0-2 years): During the first two years of life, children experience rapid growth, often doubling their birth weight by six months and tripling it by their first birthday. Height also increases dramatically, with infants growing about 10 inches in the first year. After this initial surge, growth begins to slow, though children typically continue to grow at a steady pace.

- Early Childhood (2-5 years): During the toddler and preschool years, children continue to grow steadily, gaining about 2-3 inches in height and 4-6 pounds in weight each year. Growth during this period is slower compared to infancy but remains consistent.

- Middle Childhood (6-12 years): Children in this age group grow at a steady, predictable rate, typically gaining about 2 inches in height and 5-7 pounds in weight per year. This is a period of gradual growth that prepares the body for the rapid changes of puberty.

- Adolescence (13-18 years): The onset of puberty triggers a major growth spurt, with boys and girls experiencing rapid increases in height, weight, and muscle mass. Girls typically begin puberty earlier than boys and may have completed most of their height growth by age 16, while boys continue to grow into their late teens. Adolescents can grow several inches per year during the peak of their growth spurts.

3. The Role of Genetics:

Genetics plays a significant role in determining a child's ultimate height and growth pattern. Children often follow similar growth patterns to their parents, with tall parents tending to have taller children and shorter parents tending to have shorter children. However, genetic potential must be supported by proper nutrition, a healthy environment, and overall good health for children to reach their full growth potential.

While genetics primarily dictates height and growth patterns, other factors—such as the timing of puberty—also have a genetic basis. Children who experience early or delayed puberty may exhibit different growth patterns compared to their peers. For example, early bloomers may hit their growth spurt sooner but stop growing earlier, while late bloomers may continue growing into their late teens or early twenties.

Recognizing Growth Disorders Early

Identifying growth disorders as early as possible is crucial for ensuring that children receive appropriate interventions to support their development. Growth disorders can manifest in various ways, including short stature, delayed weight gain, or abnormal body proportions. Some growth disorders are caused by hormonal imbalances or genetic conditions, while others may result from nutritional deficiencies or chronic illnesses.

SIGNS OF GROWTH DELAY

1. Short Stature:

Short stature is a common indicator of a potential growth disorder, though it is important to differentiate between children who are simply shorter than their peers due to genetic factors and those with pathological causes of short stature. Children are considered to have short stature if their height falls below the 5th percentile on growth charts for their age and sex. While some children may naturally be small due to familial short stature, others may have underlying conditions that affect their growth.

Common causes of short stature include:
- Constitutional growth delay: This is a common and non-pathological cause of short stature, where children grow at a slower rate but eventually catch up to their peers. Children with constitutional growth delay often have a family history of late bloomers and may enter puberty later than average, but they typically reach normal adult height.
- Growth hormone deficiency: A lack of sufficient growth hormone production by the pituitary gland can lead to short stature and delayed growth. Growth hormone deficiency may be present from birth or develop later in childhood. Children with this condition often have normal body proportions but grow more slowly than their peers.
- Turner syndrome: A genetic condition that affects girls, Turner syndrome results from the complete or partial absence of one X chromosome. Girls with Turner syndrome are often shorter than average and may experience delayed or absent puberty.
- Chronic illnesses: Conditions such as cystic fibrosis, chronic kidney disease, or gastrointestinal disorders can interfere with a child's ability to absorb nutrients or maintain adequate energy levels, leading to slowed growth or short stature.

2. Delayed Weight Gain:

Weight gain is another important indicator of growth and development. Children who fail to gain weight at the expected rate may be diagnosed with failure to thrive, a condition that can result from inadequate calorie intake, malabsorption of nutrients, or underlying medical conditions. Failure to thrive is most commonly seen in infants and toddlers, but it can also affect older children.

Signs of delayed weight gain include:
- A child consistently falling below the 5th percentile for weight on growth charts
- A child whose weight gain slows or plateaus over time
- A child who is significantly smaller than their peers in both weight and height

Possible causes of delayed weight gain include malnutrition, feeding difficulties, food allergies, gastrointestinal disorders (such as celiac disease), or metabolic conditions. Early intervention is essential for addressing the root cause of delayed weight gain and ensuring that children receive the nutrients they need to grow and develop properly.

3. Delayed or Absent Puberty:
The timing of puberty varies widely, but significant delays in puberty may be a sign of a growth disorder or hormonal imbalance. Most girls begin showing signs of puberty, such as breast development, between the ages of 8 and 13, while boys typically start puberty between the ages of 9 and 14. Delayed puberty is defined as the absence of sexual development by age 13 in girls and by age 14 in boys.

Possible causes of delayed puberty include:
- Constitutional delay of growth and puberty: This is a variation of normal growth and development, where puberty starts later than average but progresses normally once it begins. This is often seen in children with a family history of late bloomers.
- Hypogonadism: This condition occurs when the body produces insufficient sex hormones, leading to delayed or absent sexual development. Hypogonadism can result from genetic conditions, such as Klinefelter

syndrome in boys or Turner syndrome in girls, or it can be caused by disorders of the pituitary or hypothalamus.

- Chronic illnesses: Conditions that affect overall health and nutrition, such as Crohn's disease or severe asthma, can delay the onset of puberty by disrupting the body's ability to produce hormones necessary for sexual development.

Diagnosing Delayed Growth

When a child shows signs of delayed growth, it is important to consult a pediatrician or specialist for a thorough evaluation. The diagnostic process for identifying growth disorders typically involves a combination of medical history, physical examination, growth chart analysis, and diagnostic tests to determine the underlying cause of the delay.

1. Medical History and Family History:
The first step in diagnosing delayed growth is taking a detailed medical history, including information about the child's birth weight, feeding habits, illnesses, and developmental milestones. The doctor will also inquire about the family's growth patterns, as genetic factors can significantly influence height and the timing of growth spurts. If parents or siblings experienced delayed growth or late-onset puberty, the child may be following a similar pattern, which can provide important context for interpreting growth data.

2. Physical Examination:
A comprehensive physical examination helps assess the child's overall health and identify any physical signs that may indicate a growth disorder. The doctor will measure the child's height, weight, and body proportions, comparing these measurements to standard growth charts. They will also look for signs of puberty, such as breast development in girls or testicular enlargement in boys, as well as other physical characteristics associated with genetic conditions, such as webbed neck or low-set ears in Turner syndrome.

3. Growth Chart Analysis:

Tracking a child's growth over time using growth charts is one of the most effective tools for diagnosing growth delays. By plotting a child's height, weight, and BMI at regular intervals, pediatricians can identify any deviations from the expected growth patterns. A child who consistently remains within a particular growth percentile may not raise concern, but a child whose growth trajectory shows a sudden drop or flattening of the curve may be flagged for further investigation.

In some cases, the child's growth may be below the 5th percentile for height or weight, but if the child has always been in that range and is healthy overall, it may simply reflect a familial pattern. However, a consistent failure to follow a growth curve, especially when paired with other symptoms such as delayed puberty, weight loss, or chronic illness, can indicate a more serious growth disorder.

4. Bone Age Assessment:

A bone age assessment is a diagnostic tool used to evaluate the maturity of a child's bones in comparison to their chronological age. This test is typically done through an X-ray of the left hand and wrist, which allows doctors to examine the growth plates in the bones. The results provide insight into whether a child's bones are maturing at the expected rate or if there is a delay in skeletal development.

Bone age is especially useful in cases where delayed puberty or short stature is suspected. A child with delayed bone age may still have potential for additional growth, as their growth plates remain open longer than those of their peers. On the other hand, if a child's bone age is advanced relative to their chronological age, it may suggest that growth will stop earlier than expected, potentially resulting in short stature.

5. Blood Tests:

Blood tests are often used to diagnose hormonal imbalances or deficiencies that may be contributing to delayed growth. These tests can measure levels of growth hormone, thyroid hormones, sex hormones (such as estrogen and testosterone), and other hormones produced by the pituitary and adrenal glands.

- Growth hormone deficiency (GHD) can be confirmed by measuring growth hormone levels in the blood, typically after stimulating the body to release the hormone. If levels are low, this can indicate a problem with the pituitary gland's ability to produce sufficient growth hormone, leading to stunted growth.

- Thyroid disorders such as hypothyroidism can also cause delayed growth, as thyroid hormones play a key role in regulating metabolism and growth. Blood tests that measure levels of thyroid-stimulating hormone (TSH) and thyroxine (T4) can help identify thyroid-related growth issues.

- Sex hormone levels are measured to assess the onset and progression of puberty. Low levels of testosterone in boys or estrogen in girls may indicate delayed puberty, which can affect overall growth patterns.

6. Genetic Testing:
If a genetic condition is suspected as the cause of delayed growth, genetic testing may be recommended. This testing can identify chromosomal abnormalities or mutations that are associated with specific growth disorders, such as Turner syndrome, Klinefelter syndrome, or Noonan syndrome. Genetic testing can provide a definitive diagnosis in cases where growth delay is linked to a genetic cause, allowing for more targeted treatment and management.

7. Other Diagnostic Tests:
In some cases, additional tests such as imaging studies (MRI or CT scans) may be used to evaluate the structure and function of the pituitary gland or hypothalamus, especially if a tumor or structural abnormality is suspected

as the cause of growth hormone deficiency. Gastrointestinal tests may be performed if malabsorption or chronic illness is affecting the child's ability to absorb nutrients, leading to delayed growth.

Treatment and Management of Growth Disorders

Once a diagnosis has been made, the treatment and management of growth disorders will depend on the underlying cause. For many children, early intervention is key to improving growth outcomes and helping them reach their full height potential.

1. Growth Hormone Therapy:
Children diagnosed with growth hormone deficiency may benefit from growth hormone therapy, which involves daily injections of synthetic growth hormone to stimulate growth. Growth hormone therapy is most effective when started early in childhood, as it allows the child to make up for lost growth and achieve a height closer to their genetic potential. The therapy is typically continued until the child reaches their final adult height, which is determined by the closure of the growth plates in the bones.

2. Hormone Replacement Therapy:
For children with conditions such as Turner syndrome or hypogonadism, hormone replacement therapy may be necessary to initiate or support puberty. In girls with Turner syndrome, estrogen replacement therapy is often started during adolescence to promote breast development and support overall growth. Boys with delayed puberty or testosterone deficiency may receive testosterone replacement therapy to stimulate the development of secondary sexual characteristics and promote muscle and bone growth.

3. Nutritional Support:
For children with delayed growth due to malnutrition or malabsorption, improving nutrition is a critical part of treatment. This may involve working with a dietitian to ensure the child receives adequate calories, protein,

vitamins, and minerals to support healthy growth. In cases where underlying conditions such as celiac disease or Crohn's disease are affecting nutrient absorption, managing the condition through medication or dietary changes can help improve growth outcomes.

4. Surgical Interventions:

In some cases, surgical interventions may be necessary to address the underlying cause of growth delays. For example, children with obstructive sleep apnea caused by enlarged tonsils or adenoids may require surgery to remove these tissues, which can improve sleep quality and allow for normal growth. Children with skeletal dysplasias, such as achondroplasia, may require orthopedic surgery to correct limb deformities or improve mobility.

5. Ongoing Monitoring and Support:

Children with growth disorders require ongoing monitoring to track their progress and adjust treatment as needed. Regular follow-up visits with a pediatric endocrinologist or other specialists are important for assessing the effectiveness of treatment and ensuring that growth milestones are being met. Emotional and psychological support is also crucial, as children with growth delays may experience self-esteem issues or social challenges related to their height or appearance.

In summary, recognizing the signs of growth delay and diagnosing underlying growth disorders early are essential for ensuring that children receive the appropriate interventions to support their development. By understanding what constitutes normal growth, identifying early warning signs, and utilizing diagnostic tools such as growth charts, bone age assessments, and hormone testing, healthcare providers can help children achieve their full growth potential and address any underlying health concerns that may be affecting their development.

Medical Conditions Affecting Growth

Growth in children and adolescents is a complex process that is influenced by a variety of factors, including genetics, nutrition, hormones, and overall health. While many children grow and develop without any significant issues, some experience growth disorders due to underlying medical conditions. These conditions can stem from hormonal imbalances, chronic illnesses, or other physiological disruptions that affect normal growth patterns. Understanding the medical conditions that affect growth, along with their treatment options, is essential for ensuring that children who experience growth delays receive appropriate care and interventions.

Hormonal Imbalances and Growth Disorders

Hormones play a central role in regulating growth, particularly during childhood and adolescence. Hormonal imbalances, whether due to genetic conditions or dysfunction of the endocrine system, can significantly impact growth patterns, leading to delayed growth, short stature, or excessive growth in some cases. The key hormones involved in growth include growth hormone, thyroid hormones, and sex hormones, and disruptions in their production or action can cause a variety of growth disorders.

1. Growth Hormone Deficiency (GHD):
   Growth hormone deficiency is one of the most common hormonal causes of growth disorders. Growth hormone (GH) is produced by the pituitary

gland, a small gland located at the base of the brain. GH stimulates growth in almost all tissues of the body, including bones, muscles, and organs. When the body does not produce enough GH, or if there is a problem with the body's ability to respond to GH, growth can be stunted.

Children with GHD typically experience slow or delayed growth, resulting in short stature compared to their peers. While their body proportions are normal, they may fail to reach the expected height for their age. Other symptoms of GHD may include immature facial features, a higher-pitched voice, and delayed puberty in some cases.

The causes of GHD can be congenital, meaning a child is born with the deficiency due to genetic mutations or developmental issues affecting the pituitary gland. GHD can also be acquired later in childhood, often due to brain tumors, radiation therapy, trauma, or infections that damage the pituitary gland. In many cases, the exact cause of GHD is unknown (idiopathic GHD).

Diagnosis: Diagnosis of GHD typically involves growth chart analysis to track a child's growth pattern over time, blood tests to measure GH levels, and stimulation tests that evaluate the body's ability to produce growth hormone in response to specific stimuli. Imaging studies, such as MRI, may also be used to assess the structure of the pituitary gland and rule out tumors or other abnormalities.

Treatment: Growth hormone therapy is the standard treatment for GHD. This involves regular injections of synthetic growth hormone to compensate for the body's deficiency. GH therapy is most effective when started early and continued through adolescence. With consistent treatment, many children with GHD can reach a height that is within the normal range for their age and genetic potential.

2. Hypothyroidism:

The thyroid gland produces hormones that are critical for regulating metabolism and supporting growth. Hypothyroidism occurs when the thyroid gland produces too little thyroid hormone (thyroxine or T4). This can lead to slowed growth, developmental delays, and a range of other health issues, including fatigue, weight gain, and cognitive difficulties.

In children, hypothyroidism can result in congenital hypothyroidism, which is present at birth, or acquired hypothyroidism, which develops later. Congenital hypothyroidism is typically diagnosed through newborn screening, allowing for early treatment. Acquired hypothyroidism may develop during childhood or adolescence and is often caused by autoimmune diseases, such as Hashimoto's thyroiditis, where the immune system attacks the thyroid gland.

Diagnosis: Hypothyroidism is diagnosed through blood tests that measure levels of thyroid-stimulating hormone (TSH) and thyroxine (T4). High levels of TSH and low levels of T4 are indicative of hypothyroidism, as the pituitary gland tries to stimulate the underactive thyroid to produce more hormone.

Treatment: The primary treatment for hypothyroidism is thyroid hormone replacement therapy, typically in the form of levothyroxine, a synthetic version of T4. This medication helps restore normal thyroid hormone levels, promoting proper growth and development. Children with hypothyroidism who receive timely and appropriate treatment can achieve normal growth and avoid developmental delays.

3. Precocious Puberty:

Precocious puberty is a condition in which a child begins puberty at an abnormally early age—before age 8 in girls and age 9 in boys. This early onset of puberty can lead to an accelerated growth spurt, but it often results in early closure of the growth plates in the bones, causing the child to stop growing prematurely and leading to short stature in adulthood.

Precocious puberty can be caused by early activation of the hypothalamic-pituitary-gonadal axis, which regulates the production of sex hormones. It can also result from brain abnormalities, such as tumors or trauma, or conditions that cause the ovaries or testes to produce sex hormones independently of the brain's signals.

Diagnosis: Diagnosis of precocious puberty involves a physical examination to assess signs of early sexual development (such as breast development in girls or testicular enlargement in boys), along with hormone tests to measure levels of sex hormones such as estrogen and testosterone. Imaging studies like MRI or CT scans may be used to assess the brain for tumors or other abnormalities that could be triggering early puberty. Bone age assessments through X-rays are also useful for determining if the bones are maturing too quickly.

Treatment: The treatment for precocious puberty depends on its underlying cause. In many cases, the condition is treated with gonadotropin-releasing hormone (GnRH) analogs, which temporarily halt the progression of puberty by suppressing the production of sex hormones. This allows the child to continue growing without prematurely closing the growth plates, helping to preserve their final adult height. Once the treatment is stopped, puberty resumes normally. For children whose precocious puberty is caused by brain abnormalities, such as tumors, surgical intervention may be necessary.

4. Hypogonadism:

Hypogonadism is a condition in which the sex glands (testes in boys and ovaries in girls) produce little or no sex hormones, leading to delayed or absent puberty. This hormonal deficiency can also impair normal growth and development, as the body relies on sex hormones like testosterone and estrogen to trigger the growth spurt associated with puberty.

Hypogonadism can be primary, meaning the problem lies with the sex glands themselves, or secondary, meaning the issue originates in the hypothalamus or pituitary gland, which fail to signal the sex glands to produce hormones. Conditions such as Klinefelter syndrome in boys or Turner syndrome in girls can lead to hypogonadism, as can damage to the hypothalamus or pituitary gland due to tumors, trauma, or radiation.

Diagnosis: Hypogonadism is diagnosed through blood tests that measure levels of sex hormones, gonadotropins (LH and FSH), and other related hormones. Genetic testing may be necessary to identify underlying chromosomal abnormalities such as Klinefelter or Turner syndrome.

Treatment: Treatment for hypogonadism typically involves hormone replacement therapy. Boys may be prescribed testosterone, while girls may receive estrogen therapy to induce puberty and support normal growth and sexual development. Hormone replacement therapy is crucial for helping children with hypogonadism develop secondary sexual characteristics, experience normal growth spurts, and improve bone density.

Chronic Illness and Its Impact on Development

Chronic illnesses can have a profound effect on a child's growth and development. Many chronic conditions disrupt the body's ability to absorb nutrients, produce hormones, or maintain overall health, leading to stunted growth, delayed puberty, or other developmental issues. The most common chronic illnesses that impact growth include gastrointestinal disorders, renal (kidney) disease, and conditions like cystic fibrosis.

1. Gastrointestinal Disorders:
   Conditions such as celiac disease, Crohn's disease, and ulcerative colitis can impair a child's ability to absorb nutrients from food, leading to malnutrition and delayed growth. These autoimmune or inflammatory conditions damage the lining of the gastrointestinal tract, reducing the absorption of essential vitamins, minerals, and calories that are critical for growth.

MEDICAL CONDITIONS AFFECTING GROWTH

- Celiac disease is an autoimmune disorder in which the consumption of gluten triggers an immune response that damages the small intestine. Children with untreated celiac disease often experience poor growth, delayed puberty, and weight loss due to malabsorption.

- Crohn's disease and ulcerative colitis are forms of inflammatory bowel disease (IBD) that cause chronic inflammation in the gastrointestinal tract. This inflammation can interfere with nutrient absorption and lead to stunted growth and delayed puberty in children.

Diagnosis: Gastrointestinal disorders are diagnosed through a combination of blood tests (to check for nutrient deficiencies and inflammation), stool tests, imaging studies, and endoscopic procedures to directly examine the gastrointestinal tract. In the case of celiac disease, specific antibodies are measured in the blood, and a biopsy of the small intestine may confirm the diagnosis.

Treatment: Managing the underlying gastrointestinal condition is essential for restoring normal growth. For children with celiac disease, adhering to a strict gluten-free diet allows the intestine to heal, improving nutrient absorption and promoting growth. For children with IBD, treatments may include anti-inflammatory medications, immunosuppressants, or biologic therapies to reduce inflammation and restore normal nutrient absorption.

2. Chronic Kidney Disease (CKD):

Chronic kidney disease can significantly impair growth in children by disrupting the body's ability to maintain proper electrolyte balance, remove waste products, and produce hormones necessary for bone growth and development. Children with CKD often experience renal osteodystrophy, a condition in which bone growth is impaired due to imbalances in calcium, phosphorus, and vitamin D levels.

CKD can also lead to growth hormone resistance, where the body produces growth hormone but the kidneys are unable to convert it into its active form. This results in stunted growth, delayed puberty, and poor overall development.

Diagnosis: CKD is diagnosed through blood and urine tests that measure kidney function, including levels of creatinine, blood urea nitrogen (BUN), and electrolytes. Imaging studies such as ultrasound or CT scans may be used to assess the structure and function of the kidneys.

Treatment: Treatment for CKD focuses on managing the underlying kidney disease and addressing any complications that may arise, such as electrolyte imbalances or growth hormone resistance. In some cases, growth hormone therapy may be recommended for children with CKD to help promote normal growth. Additionally, managing the child's diet to control fluid and electrolyte levels, as well as administering vitamin D supplements, can support bone health and development.

3. Cystic Fibrosis (CF):
Cystic fibrosis is a genetic disorder that primarily affects the lungs and digestive system. In children with CF, the body produces thick, sticky mucus that can block the airways and interfere with the pancreas's ability to secrete digestive enzymes. This impairs nutrient absorption, leading to malnutrition and stunted growth.

Children with CF often experience failure to thrive, delayed puberty, and reduced bone density due to chronic malnutrition and the body's inability to properly digest and absorb fats, proteins, and fat-soluble vitamins (such as vitamins A, D, E, and K).

Diagnosis: Cystic fibrosis is typically diagnosed through newborn screening, which includes a blood test to check for elevated levels of immunoreactive trypsinogen (IRT), a marker of CF. A sweat test, which measures the concentration of salt in sweat, is the definitive diagnostic test for CF.

Treatment: Nutritional management is a key component of treating children with CF. This may involve supplementing the child's diet with pancreatic enzymes, high-calorie formulas, and fat-soluble vitamin supplements to improve nutrient absorption. Maintaining a high-calorie diet is essential for supporting growth in children with CF, as their bodies require more energy to fight infections and maintain normal development. In some cases, growth hormone therapy may also be used to enhance growth.

Treatment Options for Growth Issues

When growth issues are identified, a comprehensive treatment plan is developed based on the underlying cause. Early diagnosis and intervention are critical for helping children with growth disorders achieve their full potential. Treatment strategies may include hormone therapies, nutritional interventions, and management of chronic illnesses to address both the root cause of the growth delay and the symptoms associated with it.

1. Hormone Therapy:
   - Growth Hormone Therapy: For children with growth hormone deficiency, growth hormone therapy is a common and effective treatment. Administered through daily injections, this therapy can stimulate growth in children who have insufficient levels of natural growth hormone. It is especially effective when started early and continued until the child reaches their final adult height.
   - Thyroid Hormone Replacement: Children with hypothyroidism are treated with levothyroxine, a synthetic form of thyroid hormone. This medication helps restore normal metabolism and supports growth by ensuring the body has adequate levels of thyroid hormone.
   - Sex Hormone Therapy: For children with delayed puberty due to hypogonadism or other hormonal deficiencies, sex hormone replacement therapy (testosterone for boys and estrogen for girls) can induce puberty and support normal growth and development.

2. Nutritional Interventions:

Nutritional support is critical for children with growth delays due to malnutrition or chronic illnesses. This may involve:

- High-calorie diets for children with cystic fibrosis or other conditions that increase their energy needs.
- Gluten-free diets for children with celiac disease to allow proper nutrient absorption.
- Supplementation of vitamins and minerals such as calcium, vitamin D, or iron for children with deficiencies that are affecting their growth.

3. Management of Chronic Illnesses:

Treating the underlying chronic illness is essential for improving growth outcomes. This may involve:

- Medications to control inflammation in children with Crohn's disease or ulcerative colitis.
- Dialysis or kidney transplantation for children with advanced chronic kidney disease.
- Antibiotic therapies and airway clearance techniques for children with cystic fibrosis to prevent lung infections and improve overall health.

4. Psychological and Emotional Support:

In addition to physical treatments, children with growth disorders often benefit from psychological and emotional support to address issues related to self-esteem, body image, and social interactions. Counseling or therapy can help children cope with the challenges of living with a growth disorder and improve their overall quality of life.

In conclusion, a variety of medical conditions can significantly impact a child's growth and development, but early detection and appropriate treatment can make a profound difference in helping children achieve their full growth potential. By addressing hormonal imbalances, managing chronic illnesses, and providing comprehensive treatment options, healthcare providers can support children in overcoming growth challenges and leading healthy, fulfilling lives.

When to See a Doctor

Growth is a critical aspect of childhood and adolescence, and while most children follow a predictable pattern of development, there are times when growth deviations may signal underlying health concerns. Growth charts are invaluable tools used by healthcare providers to monitor a child's growth and development over time, but parents also play a crucial role in observing their child's growth patterns and identifying potential issues. Knowing when to seek medical advice and understanding how to navigate pediatric consultations are essential for ensuring that any concerns about growth are addressed promptly and effectively. This chapter explores the importance of growth charts, when to seek medical advice, and how to make the most of pediatric consultations.

Growth Charts and Monitoring

Growth charts are standardized tools used by pediatricians to track a child's growth over time. These charts compare a child's height, weight, and head circumference (for infants) with the growth patterns of a large population of children of the same age and sex. By plotting a child's growth measurements on these charts during regular checkups, healthcare providers can assess whether the child is growing at a normal rate, identify any deviations from expected growth patterns, and monitor changes over time.

1. How Growth Charts Work:
   Growth charts use percentiles to compare a child's measurements with

those of other children. For example, if a child's height is in the 75th percentile, this means that they are taller than 75% of children their age and shorter than 25%. The same applies to weight and body mass index (BMI), which is calculated based on height and weight. Percentiles provide a simple way to understand how a child's growth compares to standardized norms.

Pediatricians typically use different growth charts depending on the age and sex of the child. For infants and toddlers up to age 2, the World Health Organization (WHO) growth charts are commonly used. For children and adolescents aged 2 to 20, the Centers for Disease Control and Prevention (CDC) growth charts are more appropriate. These charts track height, weight, BMI, and head circumference (in infants) over time.

2. The Importance of Regular Monitoring:

Monitoring a child's growth over time is essential because growth patterns can provide insight into overall health. Regular monitoring allows healthcare providers to detect growth problems early and intervene before they become more serious. For example, a child who consistently follows a particular growth curve, even if they are smaller or larger than average, is likely growing normally for their genetic potential. However, a child whose growth deviates from their established curve—such as a sudden drop in height percentile or a plateau in weight gain—may be experiencing growth delays that require further investigation.

Head Circumference for Infants: For infants, monitoring head circumference is particularly important, as rapid brain growth occurs during the first two years of life. An abnormally small or large head circumference can signal developmental issues or medical conditions that affect brain growth.

Body Mass Index (BMI): As children grow older, BMI becomes an important measure for evaluating whether a child's weight is appropriate for their height. A high or low BMI can indicate potential concerns, such as obesity or undernutrition, that may affect overall health and growth. Regular monitoring of BMI can help identify children at risk for weight-related health issues, allowing for early intervention.

3. Interpreting Growth Patterns:

It's important to understand that growth patterns vary between children and that short stature or being overweight does not always indicate a health problem. Familial traits play a significant role in determining a child's height and body composition. For example, children of short parents are likely to be shorter than average, while children of tall parents tend to be taller.

However, certain patterns in growth charts may indicate an underlying issue:

- Failure to Thrive: If a child consistently fails to gain weight or height at the expected rate, they may be diagnosed with failure to thrive. This condition can be caused by malnutrition, chronic illness, or other underlying medical issues.

- Growth Spurts and Plateaus: During childhood and adolescence, growth is not always linear. Children often experience growth spurts, particularly during puberty, followed by periods of slower growth. However, a prolonged plateau in growth or a sudden drop in growth percentile may warrant further investigation.

- Rapid Weight Gain or Loss: Significant changes in weight, particularly rapid weight gain, can indicate potential health problems such as endocrine disorders, obesity, or metabolic issues. Similarly, unexplained weight loss may signal malnutrition, gastrointestinal issues, or chronic illness.

By regularly monitoring a child's growth and interpreting their growth charts, healthcare providers and parents can work together to ensure that growth progresses as expected. When deviations occur, they can be addressed early to prevent long-term health consequences.

Knowing When to Seek Medical Advice

While variations in growth are often normal, there are certain signs and symptoms that should prompt parents to seek medical advice. These signs may indicate that a child is experiencing growth delays or other health issues that need further evaluation.

1. When Growth Deviates from the Expected Pattern:
If a child's growth suddenly deviates from their expected pattern—such as a significant drop in height or weight percentile—it's important to seek medical advice. While short stature or slow growth may be normal for some children, particularly if it runs in the family, a sudden or dramatic change in growth trajectory can signal an underlying issue. In particular, if a child stops growing or gains weight excessively over a short period, this warrants further investigation.

Examples of growth-related concerns include:
- A child who is shorter than 95% of their peers (below the 5th percentile for height) and shows no signs of catching up with age.
- A child whose weight has plateaued or decreased, leading to concerns about malnutrition or an underlying illness.
- A child who experiences delayed puberty, especially if they are beyond the normal age range for the onset of puberty (typically by age 13 for girls and 14 for boys).

2. Signs of Delayed or Early Puberty:
The timing of puberty varies between individuals, but significant delays or early onset can indicate hormonal imbalances or other medical conditions. Girls who have not shown any signs of breast development by age 13, or boys who have not experienced testicular enlargement by age 14, may have delayed puberty, which can impact overall growth and development. On the other hand, children who begin puberty before age 8 in girls or age 9 in boys may be experiencing precocious puberty, which can lead to early closure of growth plates and short stature if left untreated.

Early Puberty: Early puberty can sometimes be associated with conditions like precocious puberty, which causes the body to mature faster than usual. If puberty starts too early, it may result in short stature because growth plates close prematurely. If a child shows early signs of puberty, such as breast development, pubic hair, or rapid growth, it is essential to consult a doctor.

3. Unexplained Symptoms Alongside Growth Issues:

Growth concerns can sometimes be accompanied by other symptoms that indicate a more serious underlying health problem. If a child exhibits any of the following symptoms in addition to abnormal growth patterns, seeking medical advice is important:

- Fatigue, lethargy, or weakness that doesn't improve with rest.
- Gastrointestinal issues such as chronic diarrhea, vomiting, or difficulty swallowing, which could indicate malabsorption or food intolerance affecting growth.
- Frequent infections or illness, suggesting a compromised immune system that may be affecting the child's ability to grow.
- Developmental delays in motor skills, speech, or cognitive function, which may point to broader developmental issues that affect growth.

4. Concerns About Obesity or Underweight:

Growth concerns aren't limited to height delays—weight issues can also be a sign of underlying problems. If a child is gaining weight excessively and their BMI is above the 85th percentile for their age and sex, this may indicate obesity, which can have long-term consequences for health, including an increased risk for diabetes, cardiovascular disease, and metabolic disorders. On the other hand, children who are underweight and below the 5th percentile for BMI may be at risk for malnutrition, delayed growth, or chronic illness.

Parents should seek medical advice if they are concerned about their child's weight, whether due to obesity or being underweight, to rule out any underlying conditions and ensure appropriate interventions are provided.

WHEN TO SEE A DOCTOR

Navigating Pediatric Consultations

When growth concerns arise, seeking medical advice from a pediatrician or specialist is the first step in addressing potential issues. Navigating pediatric consultations effectively can help ensure that any growth-related problems are diagnosed and managed in a timely manner. Here's how parents can make the most of their child's medical appointments.

1. Preparing for the Appointment:
　Before attending a consultation, it's helpful to gather any relevant information about the child's growth and health history. This might include:

　- Growth records: If parents have kept track of their child's height, weight, and head circumference at home, bringing this data to the appointment can provide valuable context for the pediatrician. Even if the doctor already has growth chart data from previous appointments, additional information can offer a more complete picture of the child's growth over time.

　- Family history: Growth patterns often run in families, so knowing whether siblings or parents experienced delayed or early growth can provide insights for the doctor.

　- List of symptoms: If the child is experiencing additional symptoms, such as fatigue, gastrointestinal issues, or developmental delays, having a list of these symptoms and when they started can help guide the consultation.

　- Medications: Parents should inform the doctor about any medications or supplements their child is taking, as some medications can affect growth.

2. Questions to Ask the Pediatrician:
　Parents should feel comfortable asking questions during the consultation to better understand their child's growth patterns and any potential concerns. Some questions to consider include:

　- Is my child's growth pattern within the normal range for their age and sex?

　- What could be causing the deviation in my child's growth curve?

　- Are there any medical tests or evaluations needed to better understand

my child's growth delay or rapid weight gain/loss?
   - What should we expect in terms of growth in the coming months or years?
   - Are there any lifestyle, nutritional, or medical interventions that can help my child's growth?
   - Should we consider seeing a specialist, such as a pediatric endocrinologist or gastroenterologist, for further evaluation?

These questions can help parents gain a clearer understanding of their child's growth and ensure they are involved in decision-making about any necessary interventions or treatments.

3. What to Expect During the Appointment:
   During the consultation, the pediatrician will perform a thorough physical examination, review the child's growth chart, and discuss any concerns the parents have noticed. The doctor will take detailed measurements of the child's height, weight, and BMI, and compare these values to standard growth charts.

If the pediatrician identifies concerns about the child's growth, they may recommend additional diagnostic tests, such as:
   - Blood tests to measure hormone levels (such as growth hormone, thyroid hormones, or sex hormones) and check for nutritional deficiencies.
   - Bone age X-rays to assess whether the child's bones are maturing appropriately.
   - Genetic testing to identify potential chromosomal abnormalities that could be affecting growth.
   - Imaging studies such as MRI or CT scans to examine the pituitary gland or other parts of the brain involved in regulating growth.

In some cases, the pediatrician may refer the child to a specialist, such as a pediatric endocrinologist, if a hormonal imbalance or growth disorder is suspected.

4. Following Up After the Appointment:

After the initial consultation, it's important to follow the pediatrician's recommendations and schedule follow-up appointments as needed. Regular monitoring of the child's growth is essential to track progress and assess whether any interventions are working. If the pediatrician prescribed growth hormone therapy, medication, or dietary changes, parents should monitor how their child responds and report any changes or side effects to the doctor.

Parents should also stay proactive about their child's overall health, including maintaining a healthy diet, encouraging regular physical activity, and ensuring the child gets enough sleep—all of which are important for supporting growth and development.

Conclusion

Monitoring a child's growth is an ongoing process that requires attention from both parents and healthcare providers. Growth charts provide valuable insights into how a child is progressing, but parents should also remain vigilant for any signs that may indicate a growth disorder or underlying health issue. Knowing when to seek medical advice is key to ensuring that growth problems are diagnosed early and treated effectively.

By working closely with pediatricians and specialists, parents can navigate the diagnostic and treatment process with confidence, ensuring that their child receives the care and support they need for healthy growth and development. Whether addressing hormonal imbalances, managing chronic illness, or implementing lifestyle changes, early intervention can make a significant difference in a child's ability to reach their full growth potential and thrive both physically and emotionally.

Conclusion

Supporting a child's healthy growth is a multifaceted process that goes beyond monitoring height and weight. It requires a holistic approach that fosters emotional well-being, builds a strong support system, and helps children navigate their developmental stages with confidence. Understanding the physical, emotional, and psychological aspects of growth helps parents, caregivers, and educators provide the necessary tools for children to thrive. By encouraging confidence, building a reliable support system, setting realistic expectations, and preparing for the future, we can ensure that children are not only physically healthy but also emotionally resilient as they progress through each stage of development.

Encouraging Confidence Through Every Stage

Confidence plays a critical role in a child's overall development, impacting their emotional and psychological well-being. As children grow and encounter changes in their bodies and minds, fostering a sense of self-assurance becomes essential. By supporting children's confidence at every stage of growth, we help them navigate challenges more effectively, establish a positive self-image, and cultivate resilience that will serve them throughout their lives.

1. Body Image and Self-Perception:
   As children enter puberty and their bodies begin to change, they may develop concerns about their appearance, particularly if they feel they do not conform to societal norms or the expectations of their peers. Boys may feel

insecure about their height or muscle development, while girls may worry about weight gain, body shape, or other aspects of their changing bodies. These concerns can undermine a child's self-confidence, leading to anxiety, body image issues, or eating disorders.

To counter these pressures, it is important to emphasize that bodies come in all shapes and sizes and that differences in growth are normal. Parents and caregivers can encourage healthy discussions about body image, helping children understand that what they see in the media often does not reflect reality. Encouraging children to focus on their strengths—whether they are physical, intellectual, or emotional—can help them develop a positive self-image that is not solely tied to their appearance.

2. Fostering Emotional Intelligence:

Emotional intelligence, the ability to recognize and manage emotions, is another crucial component of confidence. Children who are emotionally intelligent are better equipped to handle the emotional ups and downs of growth, particularly during puberty when hormonal changes can lead to mood swings and heightened sensitivity. Parents and caregivers can help foster emotional intelligence by encouraging open communication, teaching children how to identify and express their feelings, and modeling healthy emotional responses.

Teaching children coping mechanisms for dealing with stress, such as mindfulness, deep breathing, or journaling, can also help them manage emotional challenges more effectively. As children become more comfortable with their emotions, they will gain confidence in their ability to navigate difficult situations and form healthy relationships with others.

3. Encouraging Social Confidence:

Social development is another key area where confidence is crucial. Children who feel confident in their social interactions are more likely to form strong friendships, collaborate with peers, and engage in activities that help them grow both academically and personally. Encouraging participation in extracurricular activities, sports, or social clubs can help children build social skills and confidence by allowing them to explore their interests and talents in a supportive environment.

Parents and caregivers can also support social confidence by providing opportunities for children to develop leadership skills, such as allowing them to take on responsibilities at home, or by encouraging them to take initiative in group settings. Building a child's social confidence prepares them for the challenges of adolescence and adulthood, where strong interpersonal skills are essential for success.

Building a Support System

A reliable and nurturing support system is one of the most important factors in promoting healthy growth and development. Children who feel supported by their families, educators, and peers are more likely to thrive emotionally, socially, and academically. Building a support system involves not only providing a loving and stable home environment but also fostering relationships with teachers, friends, healthcare providers, and community members who can offer guidance and encouragement throughout a child's growth.

1. The Role of Family:

The family plays a central role in shaping a child's sense of security and self-worth. Parents and caregivers are often the first and most influential figures in a child's life, providing the emotional and physical support necessary for healthy development. By offering consistent love, encouragement, and understanding, families can create an environment where children feel safe to express themselves and explore their potential.

A supportive family environment also includes clear communication and realistic expectations. Parents can help children set achievable goals and celebrate their successes, fostering a sense of accomplishment and self-esteem. Open communication within the family allows children to discuss any concerns or challenges they may be facing, whether they are related to school, friendships, or their own growth and development.

2. Educational Support:

Teachers and school staff play a vital role in supporting a child's growth and development, particularly in the areas of academic achievement and social skills. Educators can offer valuable insights into a child's strengths and areas for improvement, helping parents identify any developmental concerns or areas where additional support may be needed.

Schools can also provide resources for children who may be struggling with growth-related issues, whether these are physical (such as delayed growth or early puberty) or emotional (such as body image concerns or anxiety). Guidance counselors, school nurses, and special education professionals can offer targeted support, helping children address these challenges in a healthy and constructive way.

3. Healthcare Providers:

Pediatricians and other healthcare professionals are key members of a child's support system, especially when it comes to monitoring growth and development. Regular checkups allow healthcare providers to track a child's physical progress, identify potential health concerns, and offer guidance on nutrition, exercise, and overall well-being. For children with growth disorders or other medical conditions, specialized care from endocrinologists, nutritionists, or physical therapists may be necessary.

Building a strong relationship with healthcare providers ensures that parents have access to the information and resources they need to support their child's growth. Healthcare providers can also offer advice on how to address specific concerns, such as managing growth hormone therapy, handling early or delayed puberty, or promoting healthy eating habits.

4. Peer and Community Support:

Friends and community members also contribute to a child's sense of belonging and self-worth. Positive peer relationships help children develop social skills, empathy, and cooperation, while involvement in community activities can provide additional opportunities for growth and development. Whether through sports teams, clubs, religious groups, or volunteer organizations, children who are engaged with their communities are more likely to feel connected and supported.

Encouraging children to participate in community activities can help them develop a sense of purpose and identity, while also reinforcing the values of teamwork, leadership, and mutual support.

Setting Realistic Expectations for Development

Setting realistic expectations for growth and development is crucial for both children and their parents. Every child grows at their own pace, and while growth charts and developmental milestones provide useful guidelines, it's important to remember that these are averages, not absolutes. Setting expectations that are tailored to each child's unique needs and abilities can help prevent unnecessary pressure and stress.

1. Understanding Growth Variability:

Growth and development vary significantly from one child to another. Some children may experience early growth spurts, while others may develop more gradually and reach milestones later than their peers. It's important for parents to understand that variations in growth are often normal and do not necessarily indicate a problem. For example, late bloomers may grow more slowly during early childhood but catch up during adolescence.

Parents should focus on their child's overall progress rather than comparing them to others. Celebrating small achievements and milestones, regardless of how they compare to peers, can help children feel valued for who they are

CONCLUSION

rather than how they measure up to others.

2. Managing Expectations During Puberty:

Puberty is a particularly challenging time for both children and parents, as rapid physical and emotional changes can lead to confusion, frustration, and anxiety. Setting realistic expectations for puberty involves preparing children for the changes they will experience, including growth spurts, hormonal shifts, and the development of secondary sexual characteristics.

It's also important to remind children that puberty occurs on its own timeline and that there is no "normal" age for starting or completing puberty. While some children may begin puberty as early as age 8, others may not start until age 13 or 14. Parents should offer reassurance that delayed or early puberty is usually normal and that everyone's body develops at its own pace.

3. Balancing Achievement with Support:

Encouraging children to set goals and work towards them is important for building self-confidence and resilience. However, it's equally important to balance achievement with emotional support. Children should feel that they have the freedom to pursue their interests and goals without fear of failure or criticism.

Parents can help by setting realistic goals that align with their child's abilities and by offering praise and encouragement for effort, not just outcomes. By focusing on the process of growth rather than the end result, parents can help their children develop a growth mindset, which fosters resilience and a positive attitude towards challenges.

Looking Forward: Preparing for the Next Phase

As children grow and develop, it's important to prepare them for the next phase of their lives, whether that's moving from childhood to adolescence, entering high school, or transitioning into adulthood. This preparation involves more than just physical growth—it also includes emotional, social, and cognitive development. By providing guidance and support through each transition, parents and caregivers can help children navigate new challenges and opportunities with confidence.

1. Transitioning from Childhood to Adolescence:

The transition from childhood to adolescence is one of the most significant periods of growth and change in a young person's life. During this time, children not only experience rapid physical changes but also begin to develop a stronger sense of identity, independence, and responsibility.

Preparing for this transition involves open communication about the changes that will occur during puberty, both physically and emotionally. Parents can help by normalizing the changes that come with adolescence and providing reassurance that these changes are a natural part of growing up. It's also important to encourage independence while still offering guidance and support. Parents can gradually give adolescents more responsibilities and allow them to make decisions, helping them build confidence and learn to navigate challenges on their own. Open, honest discussions about topics such as body changes, emotional swings, friendships, and peer pressure are crucial during this phase. This helps adolescents feel understood and equipped to handle the new realities of their development.

2. Navigating the Teenage Years:

The teenage years bring new social dynamics, academic pressures, and the desire for increased autonomy. Adolescents often face challenges related to peer relationships, self-identity, and the balancing act between independence and parental guidance. Preparing teenagers for these years involves fostering resilience, teaching problem-solving skills, and encouraging healthy communication.

Parents can support their teens by creating an environment where open dialogue is encouraged and by being available for guidance without being overbearing. This helps teens feel comfortable discussing concerns or uncertainties about their physical and emotional growth, social interactions, and future aspirations. It's also helpful to provide practical advice on handling peer pressure, setting boundaries in relationships, and managing academic or extracurricular responsibilities. This level of preparation helps teenagers transition through adolescence with confidence and security.

3. Preparing for Adulthood:
   As children approach the end of adolescence and prepare to enter adulthood, the focus shifts to helping them develop the skills and mindset needed for independence and self-sufficiency. This includes not only academic or career preparation but also the emotional maturity to manage relationships, self-care, and long-term personal goals.

Parents and caregivers can play a vital role in teaching life skills such as budgeting, time management, and decision-making. Additionally, guiding young adults through discussions about long-term health, including nutrition, exercise, mental health, and wellness habits, is important for laying the groundwork for a healthy adulthood. Preparing for adulthood also involves setting realistic expectations about the challenges that come with independence, such as managing stress, making career choices, or navigating complex relationships. Helping young adults develop the ability to adapt to these changes with confidence and resilience ensures that they are better prepared for the realities of adult life.

4. Lifelong Learning and Growth:
   Growth does not stop at the end of adolescence. As young people transition into adulthood, they continue to learn, grow, and evolve. Encouraging a mindset of lifelong learning is essential for fostering adaptability, creativity, and problem-solving skills. Whether through formal education, new hobbies, travel, or professional development, adults who embrace learning are better equipped to thrive in a rapidly changing world.

Fostering curiosity and a love for learning starts early in childhood and should be reinforced throughout adolescence and adulthood. Encouraging children and teens to pursue their interests, challenge themselves, and seek out new experiences helps them develop a flexible and resilient approach to life's challenges. Parents and caregivers can model this behavior by demonstrating a commitment to their own personal growth and learning.

In conclusion, supporting healthy growth extends far beyond the physical aspects of development. It encompasses emotional, social, and cognitive growth at every stage of life. By encouraging confidence, building a strong support system, setting realistic expectations, and preparing children and adolescents for the next phases of life, parents and caregivers can help ensure that young people grow into well-rounded, capable adults. Every stage of growth brings its own unique challenges and opportunities, and with the right support and guidance, children can navigate these stages with resilience, self-assurance, and a sense of purpose.

www.ingramcontent.com/pod-product-compliance
Lightning Source LLC
Chambersburg PA
CBHW050308230526
45471CB00005B/2075